roadside geology

of Colorado

Halka Chronic

MOUNTAIN PRESS PUBLISHING CO.
Missoula, Montana

1980

ROADSIDE GEOLOGY SERIES
Editorial Directors:
David Alt and Donald Hyndman

Index compiled by
Dell R. Foutz and Darlene Skipper, 1986

The paper used in this publication meets the minimum requirements of the American National Standard for Permanence of Paper for Printed Library Materials Z39.48 — 1984.

Copyright © 1980
Mountain Press Publishing Company
Ninth Printing, May 1991

Library of Congress Cataloging in Publication Data

Chronic, Halka.
    Roadside geology of Colorado.

    Bibliography: p.
    1. Geology—Colorado—Guide-books.   2. Roads—
Colorado.   I. Title.
QE91.C49   557.88   79-11148
ISBN 0-87842-105-X

MOUNTAIN PRESS PUBLISHING COMPANY
P.O. Box 2399
Missoula, MT 59806
(406) 728-1900

Colorado's colorful scenery springs from colorful geology. Its broad eastern plains are geologically and scenically different from the alpine ridges of its central mountain chains, and the rugged volcanic regions of southwestern Colorado are geologically and scenically different from the bold western deserts.

Dream Lake, a mile by trail from the Bear Lake road in Rocky Mountain National Park, lies in one of the park's many pristine rock-walled glaciated valleys. Thanks to National Park protection, this spot is as beautiful today as in 1916, when this photograph was taken. W.T. LEE PHOTO, COURTESY OF USGS

preface

*This book is designed to give to Coloradoans and Colorado visitors – especially those with little or no geologic training but with an abiding interest and sense of excitement in their surroundings – a feeling for the long prehistory of Colorado. Colorado's story, which begins not millions but **billions** of years ago, involves tremendous forces working deep within our earth. It involves rain and snow and desert wind, washing seas, and creeping tongues of ice. It involves the birth and the evolution of life, first in the sea, later on land, and still later in the air. It involves earthquakes and floods, landslides and volcanoes, and the coming of man himself.*

To avoid drowning the reader in a sea of names and facts I have deliberately told Colorado's story in simplified form, leaving out many geologic details and avoiding over-use of specialized geologic terms. But the trained geologist who comes across this book may find here at least an introduction to Colorado's geology – the skeleton and some of the flesh; the details he must seek in the geologic literature. To non-geologists – Welcome to a voyage of discovery.

In Colorado, rocks are everywhere. Clothed less often in vegetation than in wetter parts of the world, they display themselves boldly, inviting you to linger and examine. Stop often. Get out of the car and walk. Seek out a path or a trail. Take a close look at rock and fossil and mineral, and turn to this book to fit what you find into the whole. I've made a special point of pointing out features near the Rest Areas – on the Interstates the only places where non-emergency parking is permitted. Along other highways, park where you can safely do so. But please respect private property, and do keep in mind that any land that is not private belongs to you and me. Enter but do not destroy or deface or desecrate with litter. National Parks and National Monuments deserve special attention. Look and enjoy and participate in campfires and interpretive walks. These parks and monuments merit well the advice "Take only pictures, leave only footprints," so that they will remain as beautiful and as interesting for our children and our children's children.

Material in this volume comes largely from published geologic literature, much of it by members of the United States Geological Survey. To all the geologists who have mapped these plains and hiked these mountains, and shared their knowledge through the literature, I owe a debt of gratitude, as I do to the many geologists with whom I have discussed Colorado geology. To those colleagues at the USGS and elsewhere who were good enough to read and comment on the manuscript for this book, also my heartfelt thanks. National Forest, National Park, and National Monument personnel, and librarians at the University of Colorado and the USGS, are on my thank-you list as well. And to Jack Rathbone, who generously contributed many of the photos, and my daughter Emily Chronic, who devoted her artistic talents to the monotony of maps and diagrams – thank you, too.

Speaking of maps – the maps in this book are derived largely from the geologic map of Colorado published by the USGS in 1935. An updated, far more detailed map is now in press, just (at this writing) a few months away. It presents of course the newest in geologic knowledge of the state. But because that knowledge has progressed by leaps and bounds in recent years, the new map is much more complex, and I feel that for the uninitiated it is correspondingly more confusing and more difficult to interpret.

With only three exceptions, all the roadlogs in this book read from east to west or north to south. At first I tried to arrange the logs so they can be read backwards, paragraph by paragraph, by travelers going in the other direction, but my attempt was only partly successful. I apologize, therefore, to those heading east and north. You are not alone: highways in Colorado are marked with mileposts reading from south to north and from west to east, the opposite direction from these roadlogs. Mileposts on the Interstates are green and white signs on both sides of the highways. On U.S. and Colorado highways small green mile signs appear on the left if you are going toward lower mileposts, on the right if toward higher mileage numbers. Where possible I've referred geologic notations to cultural and geographic features like towns, rivers, and passes. But in Colorado's wide open spaces it has been convenient sometimes to rely on the mileposts. None of the roadlogs need relating to your car's mileage meter.

For finding your way around, use any good road map of Colorado in conjunction with this book. Most roadmaps will add interest to your travels by identifying and giving elevations for towns, peaks, and passes mentioned here.

Despite the growing worldwide movement toward use of the metric system, I've stuck with miles for horizontal distances and feet for vertical distances, merely because highways and road maps are still marked that way.

There are four types of illustrations in this book: photographs of course, maps, cross sections parallel to or at right angles to highways, and stratigraphic diagrams. The geologic maps show the age of the rock present at the surface or below soil layers. Often specific geologic names are used for specific easily recognized rock units or **formations** *(in some cases* **groups** *of formations). The same applies in the cross sections. The vertical dimensions in the cross sections are invariably exaggerated, so don't be startled if mountains seem too high or valleys too deep.*

The stratigraphic diagrams need a little explaining. They do not necessarily represent the rock layers, or **strata**, *present at any one spot, but they show diagrammatically the layers that can be seen from any one stretch of highway as you drive along. The diagrams show vertical cliff faces for rocks that* **tend** *to (but do not always) form cliffs, and sloping faces for rocks that* **tend** *to (but do not always) form slopes or "benches." Symbols are used to represent rock types such as limestone or shale or sandstone, and layers shown in color are more colorful (usually much redder) than layers shown in black. Legends for the rock symbols appear inside the cover – the same symbols have been used for all.*

Using this book, read Chapter I first. It includes a mini-course in geology, so refer to it as often as you need to. (If you're already acquainted with Colorado's geology you can skip Chapter I, except possibly as a refresher.) Before starting off down the highway, read the introductions to the chapters covering the area you'll be travelling in – plains or mountains, city environs, or western plateaus. Beyond Chapter I and the introductions to each of the other chapters, each roadlog stands alone.

Have a good trip!

Contents

ERA	PERIOD			AGE (millions of years ago)
CENOZOIC	Quaternary (Pleistocene Ice Age)			
				3
	Tertiary	EPOCHS	Pliocene	
			Miocene	12
			Oligocene	26
			Eocene	38
			Paleocene	54
				65
MESOZOIC	Cretaceous			
				135
	Jurassic			
				200
	Triassic			
				240
PALEOZOIC	Permian			
				280
	Pennsylvanian			
				325
	Mississippian			370
	Devonian			415
	Silurian			445
	Ordovician			
				515
	Cambrian			
				600
PRE CAMBRIAN	Lipalian Interval			
				? 1,000
				? 5,000

GREAT EVENTS IN COLORADO

Glaciation and concurrent erosion of high country, erosion of Colorado Piedmont.

Periodic faulting and volcanism in San Luis Valley and SW Colorado.

Miocene-Pliocene uplift: about 5000 feet of uplift to present elevations.

Erosion of new mountains with deposition of sand and gravel in intermontane valleys and on plains.

Lots of volcanic activity in San Juans and other volcanic centers. Numerous small intrusions, many of them volcanic conduits.

Deposition of oil shales in a huge western lake.

Laramide Orogeny: long and intense mountain-building creating most of the structure of the Rockies.

Marine, near-shore, and lagoon deposits which include some dinosaur remains.

Thick coal layers deposited in western swamps.

Floodplain, marsh, and dune deposits in a humid lowland climate supporting lush vegetation and lots of dinosaurs.

Continued erosion of Ancestral Rockies with deposition on broad floodplains and deltas, probably in a dry climate.

Continued erosion of Ancestral Rockies with redbed deposits spreading in sloping sheets around them.

Erosion of new mountains, with salt basins between and west of the ranges.

Colorado Orogeny: uplift of two great island ranges of Ancestral Rockies.

Deposition of marine shales containing remains of many marine animals, especially shellfish.

Widespread open sea with deposition of thick gray limestone.

Deposition of marine limestone and shale.

Marine deposition followed by erosion destroying Silurian deposits.

Deposition of limestone layers bearing some of the oldest fish remains; some mudflat deposits.

Deposition of marine sandstone and limestone as the sea crept east across denuded land.

A very long period of stability and erosion with mountains beveled to their roots.

Two or more periods of mountain-building, probably 1.5 and 2.5 billion years ago. Rocks tightly folded, partly melted, and recrystallized, with granite intrustions.

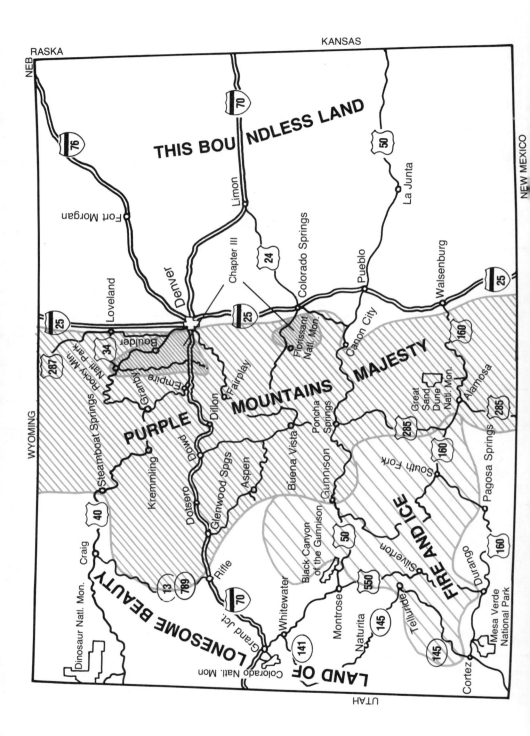

THIS BOUNDLESS LAND

PURPLE MOUNTAINS MAJESTY

FIRE AND ICE

LAND OF LONESOME BEAUTY

Chapter III

KANSAS

NEB RASKA

NEW MEXICO

WYOMING

UTAH

Fort Morgan

Limon

La Junta

Colorado Springs

Denver

Loveland

Boulder

Pueblo

Walsenburg

Rocky Mtn. Natl. Park

Granby

Empire

Fairplay

Florissant Natl. Mon.

Canon City

Great Sand Dune Natl. Mon.

Alamosa

Steamboat Springs

Dillon

Dowd

Poncha Springs

Kremmling

Glenwood Spgs

Aspen

Buena Vista

Gunnison

South Fork

Pagosa Springs

Craig

Dotsero

Rifle

Black Canyon of the Gunnison

Silverton

Durango

Dinosaur Natl. Mon.

Grand Jct.

Whitewater

Montrose

Naturita

Telluride

Mesa Verde National Park

Colorado Natl. Mon

Cortez

76

70

50

24

25

25

25

287

34

40

13

789

70

141

145

145

550

50

285

285

160

160

160

285

i
young and old
silver and gold

How old is Colorado? A thousand years? A million? Some rocks in Colorado, rocks from the northwestern mountains dated by measuring their radioactivity, tip the calendar at 2.3 **billion** years.

And 2.3 billion is a staggering figure. Half of the age of the Earth, time enough for 100,000,000 human generations. If each page of this book were to represent a single year, 2.3 billion years would build a pile of pages 27 times as high as Mt. Everest, 54 times as high as Colorado's loftiest peak, Mt. Elbert.

But Colorado as we see it today, with plains in the east, mountains through the center, basins and plateaus in the west — all of them a mile or more above the sea — goes back a paltry 20,000 years, a short life in geologic terms. Our stack of pages is a bit over six feet high.

At that time, 20,000 years ago, the high ranges were still partly clothed in glacier ice. Ridges and peaks shaped by ice and frost had begun to look like the present summits, some rounded, some sharp, some massive, some carved into delicate Gothic buttresses and spires. Lower down, canyons and valleys were flooded with ice-fed torrents white with glacial "flour" and loaded with glacier-scoured rock fragments. On the Great Plains and in the plateaus below the melting glacial tongues, lumbering mastodons and herds of bison roamed, wolves and saber-toothed tigers stalked pronghorn antelope and deer, and giant bears lumbered across river terraces or sheltered in caves that would soon be taken over by a newer arrival, man.

Colorado's geography is simple. The state is an almost perfect rectangle divided into plains, mountains, and plateau areas:

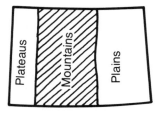

Now wiggle the lines, and insert four large valleys:

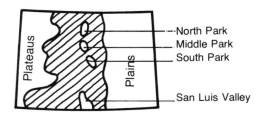

Add some volcanic mountains and a few rivers. Just about all of Colorado's rivers flow outward into neighboring states.

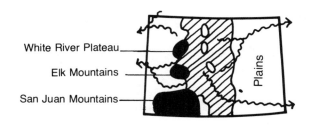

The Plains area actually divides neatly into High Plains and Piedmont.

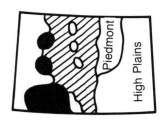

Colorado's lowest point, where the Arkansas River flows into Kansas, is 3300 feet above sea level. The plains of eastern Colorado rise from an average elevation of 4000 feet along the east border of the state to 5500 feet halfway to the mountains; then they drop off to about 5000 feet between there and the mountains. (The steps of the State Capitol in Denver are a "mile high.")

Fifty-three peaks in the state top 14,000 feet and are known affectionately as "Fourteeners." Many of them lie near or on the Continental Divide, an almost mystical dividing line between west and east, between Gulf of California (Pacific) drainage and Gulf of Mexico (Atlantic) drainage.

Continents adrift

How did the Rockies get here?

In a broad sense, the ranges of the Rocky Mountains have a history quite different from that of any other region of North America, or for that matter of the Western Hemisphere. They occupy a relatively restless geologic region that lies between the stable heart of North America, which geologists call the **craton** (a Greek word for shield), and a much less stable area

The Rocky Mountains occupy an unstable foreland between the stable North American craton and the mobile Basin-and-Range belt to the west and south.

farther west and south. According to recent geologic theory, the surface of the earth consists of a dozen or so large rigid plates looking rather like the mosaic segments of a turtle shell. The plates are about 60 miles thick, and are bordered by winding mid-ocean ridges and deep often arc-shaped oceanic trenches. At the mid-ocean ridges the plates slowly spread apart as hot lava wells up from the earth's interior to form new crust. At the trenches, plate margins are drawn under adjacent plates and eventually remelted.

The plates can be thought of as broad endless belts rolling ever so slowly, usually about an inch a year, rafting the continents along like blobs of froth on a kettle of soup. As far as we can tell, their movement is powered by convection currents deep within the earth, these in turn fueled by still deeper atomic reactions, so the whole procedure is like a slow, slow boil. Probably the rolling movement varies with time — sometimes faster, sometimes slower.

North America is on a plate that reaches from the Arctic to the Caribbean and from the East Pacific Rise, which is now close to the California coast, to the Mid-Atlantic Ridge. This plate seems to have a central weak zone along which it just isn't as rigid as elsewhere, an easily broken zone that happens to run through Colorado. During two periods of unusually intense sea-floor spreading, stresses became great enough to disrupt this weak part of the crust and break it into long, narrow blocks, some of which were lifted to form the Ancestral Rocky

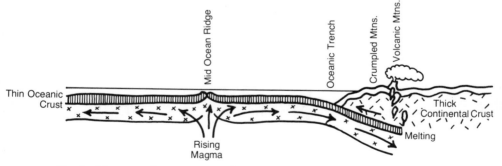

Fueled by atomic reactions deep within the earth, tremendous, powerful, slow-moving convection currents cause upwelling at mid-ocean ridges, where new crust is generated by submarine volcanic action. At the rate of about an inch a year, the oceanic crust is carried toward the continents. Where crustal plates meet, the heavier oceanic crust is pulled downward and remelted.

Mountains 300 million years ago, and then the present Rockies 60 million years ago. It probably is no coincidence that the second period of uplift, which started in California about 200 million years ago and slowly worked its way eastward, corresponded in time with the separation of North America from Europe, and therefore with the creation of the Atlantic Ocean. As the Atlantic widened, the North American Plate slid 1500 miles westward over the edge of the Pacific Plate, until the western edge of the continent itself reached the Pacific's equivalent of a mid-ocean ridge, the East Pacific Rise.

Somehow the motion in the Rocky Mountain area continued, though at a diminished pace, long after the end of the second of these mountain-building episodes or orogenies. Why it should do this is hard to understand. But apparently continued spreading along the Mid-Atlantic Ridge, and continued westward sliding of the North American Plate, caused enough inland stresses and strains from about 28 million years ago until even as recently as 5 million years ago to bow the southwest part of the continent — all of Colorado and parts of Utah, Arizona, and New Mexico — so that it now stands 5000 feet or more above sea level. With the stresses and strains involved in this upward bowing came volcanism and new movement along some of the old breaks in the crust.

The history of the plates is of course inference and educated guesswork, because we can't go back 300 million or even 10 million years, and we can't dig down 60 miles to see what's there. But the theory of drifting continents moving as parts of larger plates, plates that form at mid-ocean ridges and then sink into the crucible of the earth's interior at the trenches, is like a flash of light illuminating the uncertain darkness of past geologic thought. Most important, it establishes basic patterns that seem to fit the facts.

The Final Touch

Colorado is blessed with an exciting variety of geologic scenery, ranging from easy-to-see features of plain and plateau to intricate erosional landforms and complicated drainage patterns where streams cross rocks of varying hardness. Of particular interest to geomorphologists (geologists who study sur-

The nearly horizontal surface of the Tertiary pediment, visible in the Front Range, the Park Range, and elsewhere, was once continuous with the depositional surface of the High Plains.

face landforms) is an ancient erosion surface still apparent in the mountains. Sometimes called the Rocky Mountain Peneplain, this surface is now represented by broad rolling uplands and beveled mountain tops about 8000-9000 feet in elevation. It is quite evident in the Front Range and particularly apparent when you approach Denver and Colorado Springs from the east or northeast. It was once continuous eastward with the surface of the High Plains. Then regional uplift less than 28 million years ago increased stream flow near the mountains and caused renewed erosion along the mountain front. So now the old erosion surface — which in this book is referred to as the **Tertiary pediment** — is separated from the surface of the plains by a wide but hilly valley known as the Colorado Piedmont.

One of North America's longest rivers drains the state's western slope. Heading in high mountains on the Continental Divide, the powerful Colorado has trenched in places right through layered near-surface rocks into the ancient crystalline rocks of the continent's infrastructure, forming precipitous, narrow, spectacular canyons. Some of the Colorado's tributaries, the Yampa and the Gunnison in particular, have also carved scenic canyons. The North and South Platte Rivers, the Arkansas, and the Rio Grande, draining the eastern slope, have not made such great incisions, partly because the plains over which they flow have not been bent or lifted as strongly or as often as the western mountain flanks. For when land is lifted, the streams, be they old and slow-moving or middle-aged and swinging lazily from bank to bank, are rejuvenated and given new life and enabled to cut faster and deeper than before.

Many other weathering and erosional features vary Colorado's roadsides: joint-controlled weathering of young lava flows and ancient crystalline rocks, different degrees of erosion of hard and soft sedimentary rock layers, relatively rapid and often intricate carving of soft volcanic ash, slow, slow solution of limestone caves, and landslides that deeply scar the mountainsides. Glacial erosion shaped many mountain peaks and valleys; several small glaciers exist today in northern ranges. Wind-formed sand dunes develop in intermontane valleys, there to vie with mountain streams for possession of the land. Every turn brings a geology lesson.

Gold Is Where You Find It

Part of the fun of Colorado geology is its involvement in human history. The rocks here, and the treasures they contain, drew the first of Colorado's settlers. The rush to the Rockies began in 1859 after gold was discovered in the bed of the South Platte River near what is now Denver. Gold here is in two forms — relatively pure native gold (sometimes in ornate crystals and sometimes as shiny yellow flakes and rounded nuggets) and gold in ore minerals in combination with other elements. The first gold produced in Colorado was native gold found in stream gravels, known as placer deposits. It was mined by digging the sands and gravels of river bottoms (where gold, being heavier than other sand grains, tends to settle) and washing them, at first in gold pans, then in home-made sluices, and later in big sluice tanks in large mills or floating dredges. Miners following the placer deposits upstream sometimes "struck it rich," finding the lode from which the native gold of the streambeds came. Gold production between 1859 and 1959 centered around these towns:

Town	See route description
Central City	Denver Area
Idaho Springs	I-70 Denver to Dotsero
Cripple Creek	Colorado Springs area

Leadville	US 24 Buena Vista to Dowd
Lake City	CO 149 South Fork to Blue Mesa Res.
Creede	CO 149 South Fork to Blue Mesa Res.

Gold is still mined in Colorado, now usually as a by-product of ores of zinc and lead. The Denver Museum of Natural History and the Colorado School of Mines Geological Museum in Golden have interesting exhibits of native gold.

Several silver bonanzas encouraged fortune-seekers who arrived in Colorado in the early 1860s. As these bonanza deposits were exhausted, ores mined for lead and zinc became the principal source of silver. Notable silver-producing towns are:

Town	See route description
Aspen	CO 82 Glenwood Springs to Aspen
Central City	Denver area
Creede	CO 149 South Fork to Blue Mesa Res.
Gilman	US 24 Buena Vista to Dowd
Leadville	US 24 Buena Vista to Dowd
Silverton	US 550 Montrose to Silverton
Lake City	CO 149 South Fork to Blue Mesa Res.

Lead and zinc were discovered in Colorado during searches for gold and silver. Lead-zinc deposits are usually in places where hot mineral-rich solutions saturated old sedimentary rocks, but in the San Juan Mountains they occur in veins and fissures related to volcanoes that developed and then collapsed many millions of years ago. Principal lead and zinc towns are:

Town	See route description
Leadville	US 24 Buena Vista to Dowd
Gilman	US 24 Buena Vista to Dowd
Silverton	US 550 Montrose to Silverton

Colorado boasts the world's two largest known concentrations of molybdenum ore, at Climax and Urad-Henderson mines. (See US 24 Buena Vista to Dowd, last paragraph, and US 40 Empire to Kremmling.) The Climax mine produced more than half the world's molybdenum from 1925 through 1967.

Colorado also produces uranium, vanadium, tungsten, copper, and tin, some as by-products of molybdenum mining.

Oil has probably brought as many people to Colorado as has mining. Several oil and gas seeps were found along the mountain front shortly after the arrival of the earliest settlers. An oil well 50 feet deep drilled in 1862 near Canon City (on Oil Creek, where there was a natural seep) initiated production in the second oil field in the United States. At first, production was one barrel a day. Later, several thousand gallons of petroleum were produced there, and kerosene and lubricating oil were shipped by ox-drawn wagons all the way to Denver and Santa Fe.

The Florence oil field, about 20 miles southeast of Canon City (pronounced Canyon), was drilled in 1876. Some of the early wells in the Florence field are still producing, so this is Colorado's longest-operating field. Small quantities of oil have been pumped near Boulder since about 1900, and several old rigs can be seen northeast of town.

More recently, oil has been found a mile or more beneath the surface in the northeastern part of the state. Here it accumulated in bands of sand that were once beaches of an ancient sea. Individual "pools" are small but numerous. Oil is also found in small, rich accumulations in western Colorado, some of it more than 1½ miles below the surface.

The world's greatest known potential source of oil lies in the oil shales of western Colorado, which cover a large area north of Rifle and the Colorado River and extend well into Wyoming and Utah. The oil shales are fairly young rocks which formed in the muddy bottom of a big lake about 50 million years ago. Oily material called **kerogen,** waxy and too solid to flow through the fine pore spaces of the shale, is locked firmly into these rocks. It can be freed by mining the shale, crushing it, and heating it to high temperatures, a process likely to be at odds with environmental concerns, or by a newly developed process in which the shale is heated in place underground. But sixty years of searching for an economical way to produce oil from oil shale has not brought costs low enough to compete with world market prices.

History — The Underlying Theme

The central theme of geology — the theme that ties geology together — is history. What we know of Colorado's geologic history began 2½ billion years ago. We have no way of knowing what Colorado or North America or even the Earth as a whole looked like then. Did they look raw and new, despite their age? or old and worn and weary like the surface of the moon? We do know that oceans or at least large bodies of water existed, for some ancient rocks, altered and recrystallized though they may be now, wear the stripes associated with flat-lying water-deposited sandstone, limestone, and shale. We do know that there were no plants on the land, or animals, though primitive beginnings of life may have existed in the sea. We do know that gigantic mountain chains formed, for we can examine the distorted, twisted, granite-like rocks of their roots. We do know that volcanoes erupted, for we find unmistakable mineral groups in rocks that once were lava flows. Did an atmosphere exist? Almost certainly, for volcanoes produce gas as well as lava and ash. With equal certainty, it was not the life-supporting oxygen-rich atmosphere we know today, a product of plant metabolism.

Coming To Terms

Here at this end of geologic history — not really an end but just a way-station — geologists find they need handles for geologic time. Clocks and calendars don't do the job for units of millions and billions of years. So they have attached names to time intervals which they call **periods.** The names are usually derived from places where, early in the geology game, rocks deposited during those intervals were first described.

And just as we lump weeks into months or divide them into days, geologists lump and subdivide their time terms. The periods are lumped into **eras,** shown on the color blocks on the Geologic Time Scale (p. xii), and the names given to the eras apply to the level of development of living things preserved as fossils.

We live in the most recent era, the **Cenozoic,** whose name means "recent life," an era also often called the Age of Mammals. **Mesozoic** means "middle life," and is the Age of Reptiles, the time when dinosaurs ruled the land. **Paleozoic** means "old or ancient life," and is also called the Age of Fishes. The oldest era, once called the Azoic (without life), was at first supposed to contain no record at all of living things. We know now that life did exist during much of it, although in primitive or soft-shelled forms not easily fossilized. We now call this era **Precambrian**, and at least in its later part we could refer to it as the Age of Shell-less Invertebrates. Sometimes it is divided into Archeozoic and Proterozoic Eras but this isn't necessary in Colorado.

Only recently have we been able to pin down some exact — well, moderately exact — dates for the periods and eras. These are obtained by comparing the amounts of radioactive elements and their "daughter products" in rocks, and they are accurate only with certain types of rocks, and only within about 5%. So we round the numbers off to easy-to-remember figures and let it go at that.

The Geologic Time Scale given in the preface outlines the great events in the history of the state, with only brief suggestions of what happened elsewhere. And it lists the rock units — **formations** or **groups** — used in this guide. They, too, are named after places, places in Colorado or sometimes adjacent states where they are well developed and well exposed. Sometimes in the text you'll find references to the time in years when a certain thing happened, but more often, to save space, it's left out, and you can refer back to the Time Scale if you want the figures. The important thing is to visualize the **relative** position of the rock layers — which one is above and therefore younger, or which is below and therefore older. For according to the "law of superposition," one of geology's basic tenents, geologic history begins at the bottom, with layered rocks becoming younger upward. (There are exceptions to this "law," and this book will point out a few in Colorado.)

A few terms are in such common geologic use that you really can't get along without them. You may already have come across them in newspapers and magazines. A **fault** is a break, or a zone of more or less parallel breaks, along which movement has taken place. A **joint** is a rock fracture along which no perceptible movement has occurred. An **anticline** is an upward bend or fold in rocks, usually visible only in layered rocks

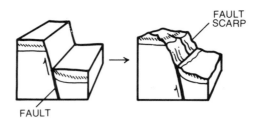

FAULT SCARP

FAULT

or **strata**. A **syncline** is, conversely, a downward bend or fold. Anticlines and synclines are not always symmetrical — they may be lopsided or **asymmetrical**.

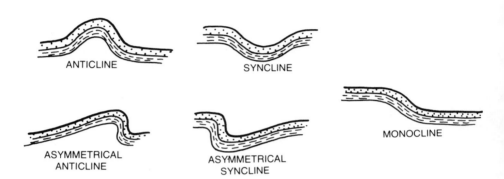

ANTICLINE SYNCLINE

ASYMMETRICAL ANTICLINE

ASYMMETRICAL SYNCLINE

MONOCLINE

Geologists use the word **orogeny** for mountain-building episodes, and they name orogenies after geographic features, usually the very mountains they create. So we get the Nevadan Orogeny producing the Sierra Nevada, or the Laramide Orogeny from the Laramie Range of Colorado and Wyoming, part of the Southern Rockies.

Other terms that are used frequently are of course the names of rocks. There are literally hundreds of kinds of rocks recognized by geologists, but for everyday use a baker's dozen will suffice — see the accompanying chart. They group into three main classes depending on their origin:

• **Igneous rocks** originate from molten rock material or **magma** that rises from as much as 200 miles below the surface. The magma may come to the surface of the earth as a volcanic or **extrusive** igneous rock or may cool slowly below the surface as **intrusive** igneous rock. Intrusive rocks are usually coarser-grained than their extrusive counterparts.

12

Class	Rock	Description
Sedimentary	Sandstone	Grains of sand cemented together
	Shale	Grains of silt and clay cemented together, usually breaking into flat slabs
	Conglomerate	Sand and pebbles deposited as gravel and then cemented together
	Limestone	A sedimentary rock composed mostly of calcite, deposited as a limy mud. Usually white or gray, often containing fossils
Igneous Extrusive	Latite	Fine-grained volcanic flows and ash composed mostly of feldspars
	Rhyolite	Light-colored very fine-grained volcanic rock, either flows or ash
	Basalt	Very fine-grained black volcanic rock, often with gas bubbles
Igneous Intrusive	Granite	Common light-colored coarse-grained rock with visible crystals of quartz and feldspar, usually peppered with black mica or hornblende
	Monzonite	Medium-grained rock (often with some larger grains) made predominantly of feldspar, with some hornblende or quartz
Metamorphic	Marble	Recrystallized limestone, often with visible calcite crystals
	Quartzite	Sandstone so tightly cemented that it breaks through the individual sand grains
	Gneiss	Banded or streaky crystalline rock formed from older granite or sandstone
	Schist	Medium-grained rock with mica grains lined up so that rock has streaky appearance and tends to split along parallel planes

Thirteen common rocks of Colorado

Sweeping diagonal cross bedding indicates that this sandstone was deposited originally in ancient sand dunes.

C.E. ERDMAN PHOTO, COURTESY OF USGS

• **Sedimentary rocks** are formed from broken or dissolved bits and pieces of other rock, loosened by frost or water or gravity and carried by wind or water or ice to be deposited as layers of fragments or as chemical precipitates. Most obvious layering in rocks is sedimentary. Usually the youngest sedimentary rocks, nearest the top of any rock sequence, are softer than the older ones, for sedimentary rock tends to harden with age. However there are exceptions.

• **Metamorphic rocks** are pre-existing rocks, either sedimentary or igneous, that have been altered by heat or pressure or chemical change. They may be scarcely altered at all or they may be changed so severely that it is difficult or impossible to figure out what they were originally.

Rocks are made of **minerals**, natural substances that have definite chemical make-ups and very often definite recognizable ways of crystallizing. A few common rock-forming minerals, and tips for recognizing them, are listed below. Rockhounds and gemologists sometimes use different terms, though they will doubtless recognize many old friends in Colorado.

• **Quartz**, a clear, hard (can't be scratched with a knife), shiny, glassy mineral that is the primary mineral in most sandstone and one of the primary minerals in granite. Quartz in rocks is commonly clear and colorless, but it may be pink (**rose quartz**), lavender (**amethyst**), snow white (**milky quartz**), or gray and clear (**smoky quartz**).

• **Feldspar**, a translucent pinkish, grayish, or whitish mineral very common in granite, recognized by its tendency to break along flat cleavage faces that catch the sunlight.

• **Mica**, a soft (scratch it with your fingernail) black or silvery mineral that can be separated easily along shiny flat paper-thin cleavage faces, very common in granite and schist. In schist, lined-up mica often imparts shiny, easily broken surfaces to the rock itself. Black mica is **biotite**; white mica is **muscovite**.

• **Calcite**, a white to light gray mineral (may be colored by impurities) making up most limestone, including "dripstone" in caves. It can't be scratched with a fingernail but can with a knife. Geologists test this mineral with dilute acid — it fizzes.

• **Hematite,** an iron mineral that is easily recognized by its deep brick red color, occurring as tiny grains giving color to sandstone and shale "redbeds." When concentrated, hematite is an ore of iron.

• **Limonite**, a dull rusty yellow iron compound often occurring in tiny grains that give an over-all mustard yellow or tan to sandstone and shale. In Colorado's mining areas limonite colors old mine dumps. When concentrated it is an ore of iron.

• **Kaolin**, a white clay mineral abundant in some shale and sandstone, formed by disintegration of feldspar. This mineral is most easily recognized by its clayey, earthy odor. (Dampen the rock with your breath before sniffing.)

• **Hornblende**, a black mineral in rod-like crystals (as distinct from flat mica crystals) common in dark igneous and metamorphic rocks.

• **Gypsum**, a soft (scratch it with a fingernail) transparent or translucent colorless mineral formed as an evaporite where sea water or salty ponds dry up. Gypsum crystals are **selenite**.

• **Pyrite**, "fool's gold," a brittle metallic brass-colored mineral sometimes mistaken for the real thing. Pyrite is a compound of iron and sulfur, and when abundant is an ore of iron.

There are many other geologic terms that will be used in this book from time to time. These are defined where first used, but it's so unlikely that you'll drive the roads in the order given here, that they are all defined again in the Glossary. Some of the most important terms are defined by sketches for quick reference.

ii
this boundless land
— plains and piedmont

The Great Plains — the nation's breadbasket. Mile on mile of wheat ripens in the sun, and tall granaries are dwarfed by taller sky. Seemingly flat as a pancake when seen from a distance, this gently rolling upland was once home to wandering Cheyenne, Kiowa, and Comanche; once it shook with thundering herds of bison.

In Colorado, part of the Great Plains called the High Plains rises gently westward from around 4000 feet at the border with Kansas and Nebraska to 5000 feet or more at its western rim. The rock beneath the sandy, pebbly soil is composed of sand, gravel, and clay that were washed off the mountains late in Tertiary time. Below this poorly consolidated rock — and there may be 700 feet of it — are older layered sedimentary rocks, including many thousand feet of thick, gray Pierre Shale deposited 70-80 million years ago in a shallow Cretaceous sea.

The history represented by these two groups of sediments — the Pierre Shale and the pebbly Tertiary gravel — spans an eventful time, for between deposition of the two the Rocky Mountains were born. Previously, during both the Paleozoic Era and the first part of the Mesozoic Era, seas had inundated the continent again and again, and the Ancestral Rockies, here for a time, were long since washed away. Then, in a quiet shallow sea that stretched from the Arctic to the Gulf of Mexico, fine gray muds of the Pierre Formation accumulated. Strange animals swam in the sea and crawled on its muddy bottom; sometimes when they died their shells or skeletons or teeth became buried in the mud and preserved as fossils.

Westward, the shore sloped gently upward, a coastal plain quite like the broad and almost featureless slope of the southern Atlantic and Gulf coasts of United States today. And as in

90 million years ago, shoreline features were migrating east as land rose in the west.

Around 50 million years ago, the Denver and Dawson Formations were deposited in a basin east of the Rockies. These formations were later eroded back to the dotted line.

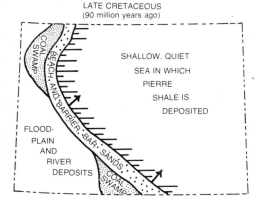

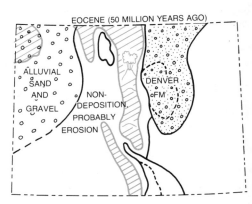

By 75 million years ago, the sea had left Colorado. An apron of debris spread outward from the rising Rockies.

20 million years ago, the mountains were nearly covered with their own debris. The entire area was then uplifted some 5000 feet.

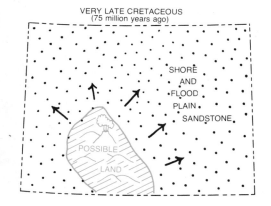

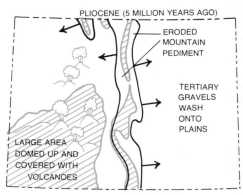

The birth of the Rockies

those lowlands, a line of beaches and lagoons and barrier bars separated land from sea. Nothing resembling a mountain was closer than western Utah, four hundred miles away.

Then slowly, slowly, around 75 million years ago, toward the end of the Mesozoic Era, parts of Colorado began to rise. The sea, with its muddy bottom, with its beaches and barrier bars and lagoons, slowly withdrew eastward across the state, leaving above the old Pierre Shale layers of beach and bar sand, lagoon mud, and plant-filled swamps that would someday turn to coal. Colorado became for a time part of a widening coastal plain. And at last, at the beginning of the Cenozoic Era (Age of Mammals), 65 million years ago, mountains began to form, apparently in response to titanic forces on faraway mid-ocean ridges.

As the mountains rose, land on either side sagged a little as if to counterbalance the mountain uplift. As the mountains grew higher and the sags grew deeper, stream gradients steepened and streams and rivers became more active. Rejuvenated, they carved into the rising hills, carrying away heavier and heavier loads of sand and silt and gravel, all of which they unceremoniously dumped east and west into the sagging basins.

Through the early part of the Cenozoic Era, for perhaps 20 million years, the mountains continued to rise and the basins next to them to sink. About 45-50 million years ago, both

Two types of erosion and deposition around isolated mountain uplifts.

mountains and basins became more stable. Erosion continued unabated, of course, until by 28 or 30 million years ago the mountains were worn down until only scattered ridges and peaks remained, half drowned in their own debris — debris that filled the basins, overflowed them, and spread like a broad sloping apron far into what is now Nebraska and Kansas. The apron was continuous westward with its erosional counterpart, a surface cut right into the rocks of the mountains themselves, the Tertiary pediment, which surrounded the remaining peaks. Such pediments and sloping debris aprons are quite common today around isolated ranges of our southwestern deserts. As you will see as you approach the Colorado Rockies, remains of the old pediment, often inappropriately called the Rocky Mountain Peneplain, can be distinguished to this day.

The sag east of the Rockies, which geologists call the Denver Basin, is no longer basin-shaped on the surface. But deep beneath the surface, layered rocks that fill it show the pattern of an oval-shaped area of subsidence. The pattern extends south to Pueblo, north into Wyoming, and eastward nearly to Nebraska. In its deepest part, not far from Denver, the Denver Basin is filled with layered sediments 13,000 feet thick, in contrast to only 6000 feet near the Nebraska line. It is an important petroleum province, and as you drive across it you will come upon scattered oil wells. Both oil and gas come from Cretaceous rocks, where they fill spaces between grains in sandstone and joints and fissures in the Pierre Shale and inter-layered beds of limestone. On the west flank of the basin, oil is produced from older rocks bent into folds near the mountain front.

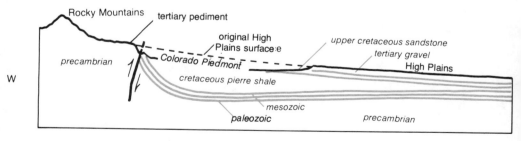

Diagrammatic cross section of High Plains and Colorado Piedmont (not to scale).

20

Sometime around 28 million years ago, after a long period of relative stability, the entire region — plains, buried basins, and mountains — began to lift into a huge broad dome, a dome that extends from the middle of Kansas to the deserts of Utah. Near the mountains, erosion slowly stripped away several thousand feet of the uppermost sediments, excavating a broad and gentle valley, the Colorado Piedmont, between what by now can be called the High Plains and the Rocky Mountains. Streams were again rejuvenated, and as they scoured and cleaned the mountain slopes they revealed the shape of the pediment carved in the hard old mountain core.

During this period of erosion the South Platte River, which like other rivers on this side of the mountains had flowed eastward, rerouted its course and began to flow north along the Piedmont to join with the Cache la Poudre River. It no doubt helped to carve out the Piedmont. Many small streams changed their courses as well. Streams on the High Plains flow east today, reflecting the original drainage across the apron of debris. But streams within the Piedmont now flow northward or southward into the South Platte or the Arkansas River. Close to the mountains, controlled by the line of upturned sediments at the mountain edge, some of these streams exactly parallel the mountain front. It is they that have stripped away the High Plains veneer and uncovered early Tertiary and Cretaceous rocks in the Piedmont.

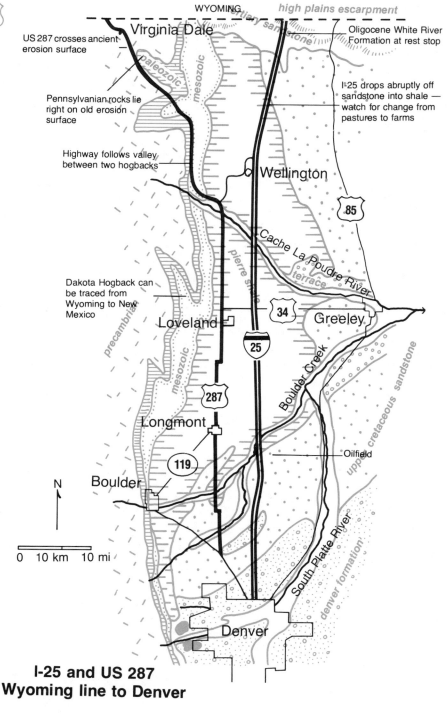

WYOMING

high plains escarpment

Virginia Dale

US 287 crosses ancient erosion surface

Oligocene White River Formation at rest stop

Pennsylvanian rocks lie right on old erosion surface

paleozoic

mesozoic

tertiary sandstone

I-25 drops abruptly off sandstone into shale — watch for change from pastures to farms

Highway follows valley between two hogbacks

Wellington

85

Cache La Poudre River

Dakota Hogback can be traced from Wyoming to New Mexico

precambrian

pierre shale

terrace

34

Loveland

Greeley

25

mesozoic

287

Boulder Creek

Longmont

119

Oilfield

upper cretaceous sandstone

N

Boulder

0   10 km   10 mi

South Platte River

denver formation

Denver

I-25 and US 287
Wyoming line to Denver

*At the edge of the High Plains, erosion has carved
rugged sandstone figures in poorly consolidated
Tertiary sedimentary rocks.*

interstate 25
wyoming — denver
(85 miles)

Just before leaving Wyoming, Interstate 25 begins a descent off the escarpment that edges the High Plains, dropping gradually into the Colorado Piedmont. For several miles it crosses Tertiary sediments, stair-stepping down through a layer of gravelly sandstone to a lower layer of very fine white to tan sandstone — a rock that erodes easily to form badlands or rock castles like those at the Rest Area at highway Mile 296. Not far east of here, in the White River Badlands of South Dakota, Nebraska, and Colorado, many fossil vertebrate skeletons have been dug out of these sediments — saber-toothed tigers, rhinoceroses, camels, little three-toed horses, giant pigs, turtles, and strange rhinoceros-like Titanotheres.

Looking north from eight or ten miles south of the Wyoming border you can see the High Plains escarpment clearly. Notice that the High Plains surface rises steadily westward in Wyoming in a long ramp-like slope known as the Gangplank. Only the uppermost peaks of the range show above it. The Gangplank made history in 1869 as the route of the nation's first transcontinental railroad — the easy way to cross the mountains.

In Colorado, there is no Gangplank. The broad valley of the Colorado Piedmont separates the High Plains from the mountains. But geologists believe that the plains once extended clear to the mountains here too, and that the South Platte, the Arkansas, and their tributaries excavated the valley of the Piedmont perhaps 5 to 10

23

million years ago, after regional uplift. Three lines of evidence support this belief: First, the High Plains rock layers slope eastward as if they were outwash from the mountains. Second, stream courses flow eastward in a way that suggests they originally flowed straight downslope from the mountains. And third, on parts of the Front Range the Tertiary erosion surface slopes in such a way that if projected it would be continuous with the High Plains to the east. Eight thousand to 10,000 feet in elevation, this surface is not at all a perfect plain, but it is nevertheless quite easily discerned.

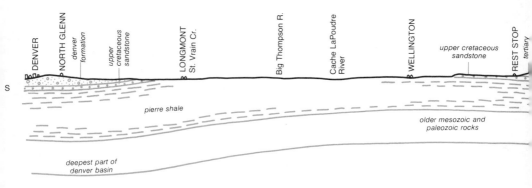

Section along I-25 from Wyoming line to Denver

At Mile 293 the highway leaves the Tertiary sediments of the High Plains and drops into the Colorado Piedmont. For about 12 miles it crosses a level, natural step of Upper Cretaceous sandstone, with a long finger-like promontory of Tertiary sediments paralleling the east side of the highway. The Cretaceous sandstone on which the highway runs was once beach and bar sand edging the retreating Cretaceous sea. Notice that it forms rather poor soil, suitable only for grazing.

At mile 281 the highway descends further — to the top of Cretaceous Pierre Shale. There is an abrupt change of land use, from grazing on the sandstone to farming on the Pierre Shale. This change is typical, and can be used in the Piedmont area to map the formations, which by and large are rarely exposed in outcrops but strongly influence the soil chemistry and texture.

The route to Denver parallels the east side of the Front Range, whose north end can be seen clearly from I-25. Like many other ranges in the Southern Rockies this one is a huge uplifted block of

24

Precambrian rocks which once had layers of sedimentary rock draping across it. Ancient Precambrian rocks, mostly granite, form the core of the range, and typically Mesozoic and Paleozoic sedimentary rocks are bent upward along its edges.

West of Fort Collins and Loveland lies one of the highest and most spectacular parts of the Front Range, Rocky Mountain National Park (see the log for U.S. 34 LOVELAND to GRANDY). The Front Range includes many 13,000-foot summits and four Fourteeners (over 14,000 feet), among them 14,255-foot Longs Peak, the prominent isolated peak in the southern part of Rocky Mountain Park. Your elevation now is about 5000 feet, so the total topographic relief here is over 9000 feet. However, **structural** relief is much greater: Precambrian rocks that form the summit of Longs Peak are 22,000 feet higher than Precambrian rocks in the deepest part of the Denver Basin. So the upward displacement of the mountains relative to the Piedmont is at least 22,000 feet! It may be still greater, but there is no way to know just how much of the old Precambrian rock has been eroded away from the top of the range.

South of Mile 250 there are good views westward to a jagged row of peaks south of Rocky Mountain Park. Ogallala, Paiute, Pawnee, Shoshone, Arapaho — they are named after Indian tribes and are known collectively as Indian Peaks. The line of their summits marks the Continental Divide, the crest separating Pacific and Atlantic drainage. These peaks, too, are composed of Precambrian rocks.

Can you pick out the hilly yet somewhat level surface below Indian Peaks and about halfway down the mountain slope? This is the Tertiary pediment, the erosional surface that formed at the foot of the mountains in mid-Tertiary time and that was once continuous with the High Plains.

For the last 40 miles before reaching Denver the highway crosses Tertiary and Cretaceous sandstone and shale. Good exposures are rare. At Mile 225 it climbs slightly onto the Denver Formation, coarse gravelly Tertiary sandstone and conglomerate that fill the center of the Denver Basin. The Interstate remains on this formation and its equivalent, the Dawson Formation, almost all the way to Colorado Springs.

Arching west through Denver, the highway follows the valley of the South Platte River. The river floodplain, now peppered with factories, roads, storehouses, and railroad yards, was deep under rushing, tumbling flood waters on June 16, 1965. The flood crested rapidly and destroyed almost everything in its path. Flood-carried debris piled up behind bridges, and bridge after bridge collapsed or washed out, effectively cutting Denver in half. The flood came from the south where the river and its tributaries drain areas that received 14 inches of rain on that fateful afternoon.

Geologic features near Denver, including Mt. Evans, Red Rocks Park and a Boulder — Estes Park — Central City loop, are discussed separately under DENVER AREA.

Rhinoceros-like Titanotheres and other strange mammals inhabited northeastern Colorado in Oligocene time. Their fossilized skeletons are exhibited at Denver Museum of Natural History.

PHOTO COURTESY OF DENVER MUSEUM OF NATURAL HISTORY

interstate 25
denver — colorado springs
(65 miles)

Between Denver and Colorado Springs I-25 crosses Tertiary rocks that fill the center of the Denver Basin, mostly light-colored tan and white and yellowish sandstone and conglomerate called, collectively, the Denver and Dawson Formations. These rocks often appear black in outcrops because black lichens grow on them. Between some layers are thin volcanic flows and volcanic ash beds, but they are hard to distinguish at Interstate speeds.

West of the route, the Front Range rises like a gigantic ocean wave to crest above 14,000 feet at Mt. Evans, Mt. Bierstadt, and Pikes Peak. Summits between these high points are hidden from I-25 by the lower Rampart Range, in the foreground.

The castle-shaped hill that gives Castle Rock its name wears a cap of hard Castle Rock Conglomerate, somewhat younger than the Denver and Dawson Formations and a good deal more resistant to erosion. This coarse pebbly rock caps similar mesas and buttes farther

Castle Rock butte gives its name to the conglomerate that caps it. This is the type section of the Castle Rock Conglomerate, the place where the rock is typically displayed and named.
HALKA CHRONIC PHOTO

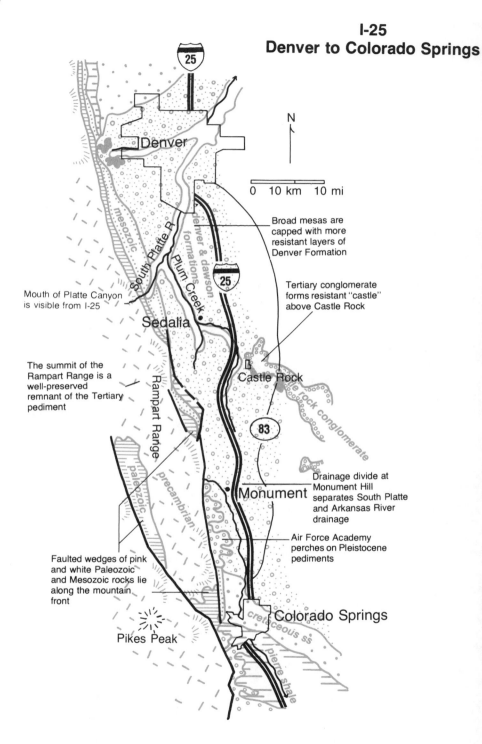

Denver

N

0   10 km   10 mi

Broad mesas are
capped with more
resistant layers of
Denver Formation

mesozoic

South Platte R

denver & dawson formations

Plum Creek

Mouth of Platte Canyon
is visible from I-25

Tertiary conglomerate
forms resistant "castle"
above Castle Rock

Sedalia

The summit of the
Rampart Range is a
well-preserved
remnant of the Tertiary
pediment

Rampart Range

paleozoic

precambrian

Castle Rock

castle rock conglomerate

83

Drainage divide at
Monument Hill
separates South Platte
and Arkansas River
drainage

Monument

Air Force Academy
perches on Pleistocene
pediments

Faulted wedges of pink
and white Paleozoic
and Mesozoic rocks lie
along the mountain
front

Colorado Springs

cretaceous ss

Pikes Peak

pierre shale

28

south. Its sand and pebbles, doubtless derived from the mountains close by to the west and southwest, may be a last remnant of a huge **alluvial fan** formed by the ancestral South Platte River, which emerged from the mountains north of Pikes Peak.

Streams reaching a sharp decrease in slope deposit part of their load in alluvial fans.

About four miles south of Castle Rock there is an excellent view of the Rampart Range. Evidence that this range is a faulted anticline can be seen between the highway and the mountains, where isolated pink rocks stand out of the pine forest. They are late Paleozoic sedimentary rocks similar to those in Red Rocks Park near Denver and the Garden of the Gods at Colorado Springs, made of fine sand, mud, and gravel washed down off the Ancestral Rocky Mountains. These, you may remember, rose in Colorado about 300 million years ago, long before the present Rockies existed. Bent upward into an almost vertical position as the present Front Range rose, they probably once extended much higher up the slope or even clear over the top of the range. Just here, there are only small wedges of these rocks. On either side of them Tertiary rocks run right up to a fault at the edge of the Precambrian Pikes Peak Granite that makes up the core of the mountains.

All along Plum Creek between Castle Rock and Monument Hill you can see flash-flood scars in the stream banks. During the afternoon of June 16, 1965, when all this area was deluged by a 14-inch cloudburst, floodwaters sweeping down Plum Creek cut into the banks and widened the channel, then joined with other streams flowing into the South Platte, and inundated the parts of Denver that were on the river's floodplain (see DENVER AREA). When they are heavily loaded with sand and silt, churning waters are so dense that they are quite able to transport huge blocks of rock, along with concrete bridges, houses, cars, or anything else in their paths.

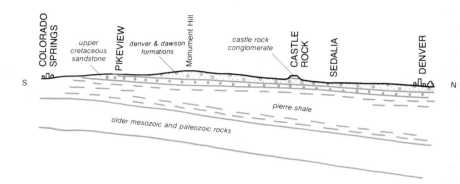

Section along I-25 from Denver to Colorado Springs

At Monument Hill, the high point between Denver and Colorado Springs (Mile 163), the highway crosses the divide between Arkansas River and Platte River drainages. A prominent white tower standing out from the mountains, an erosional left-over of the same Tertiary sandstone that is exposed in roadcuts on Monument Hill, gives the name to the town of Monument. Unlike the pink rocks farther north, which were tipped up as the mountains rose, these white rocks are nearly horizontal — they were formed **after** rather than **before** the uplift of the Rockies.

Carved by erosion, Tertiary sandstone giants line up at the edge of the forest near Monument. A hard layer of brown sandstone, resistant to erosion, forms their caps. N.H. DARTON PHOTO, COURTESY OF USGS

The Rampart Range stretches from west of Castle Rock to Colorado Springs, a lower block of the Front Range separated from it and from the foothills by faults. Its almost horizontal upper surface, like that of other parts of the Front Range, is a remnant of the Tertiary pediment eroded in the Pikes Peak Granite of the mountain core. Much younger pediments form the site of the Air Force Academy. Like their older counterpart they give evidence of long stable periods in the history of these mountains.

Scars on the mountain south of the Air Force Academy are man-made — quarries in hard Paleozoic limestone used for concrete aggregate and road material. Unfortunately, recent efforts by conservationists to stop the quarrying have been as fruitless as attempts to reseed the scars. Only time will heal these wounds.

Colorado Springs nestles in the valley at the confluence of Monument Creek and Fountain Creek, protected by hills of Cretaceous and Tertiary sandstone to the east. Older parts of the city, near these streams, are on Pierre Shale, but the growing population has overflowed westward onto pediment gravels and eastward over the sandstone hills.

Geologic features near Colorado Springs, including Pikes Peak, Cave of the Winds, and Garden of the Gods, are discussed separately under COLORADO SPRINGS AREA.

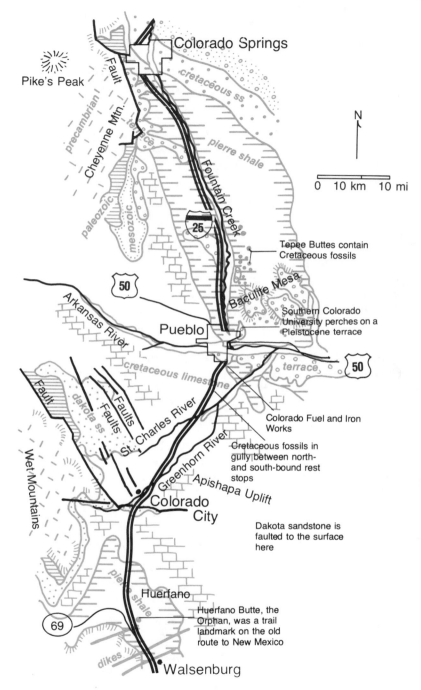

Pike's Peak

Fault

Colorado Springs

cretaceous ss

precambrian

Cheyenne Mtn. terrace

paleozoic

mesozoic

Fountain Creek

pierre shale

N

0   10 km   10 mi

25

Tepee Buttes contain
Cretaceous fossils

Baculite Mesa

50

Arkansas River

Pueblo

Southern Colorado
University perches on a
Pleistocene terrace

cretaceous limestone

terrace

50

Fault

dakota ss

Faults

Faults

St. Charles River

Greenhorn River

Colorado Fuel and Iron
Works

Cretaceous fossils in
gully between north-
and south-bound rest
stops

Wet Mountains

Apishapa Uplift

Colorado
City

Dakota sandstone is
faulted to the surface
here

pierre shale

Huerfano

69

Huerfano Butte, the
Orphan, was a trail
landmark on the old
route to New Mexico

dikes

Walsenburg

I-25
Colorado Springs to Walsenburg

interstate 25
colorado springs — walsenburg
(96 miles)

Most of Colorado Springs — particularly the older part of town — is built on Cretaceous Pierre Shale that underlies the valley of Fountain Creek between the Rampart Range and whitish, cliffy hills of Cretaceous sandstone. As far as Pueblo, Interstate 25 follows the contact between the gray marine shale and the edge of the floodplain of Fountain Creek. As a rule, Pierre Shale weathers into soil-covered slopes and is not at all well exposed. In a number of roadcuts, though, and sometimes in gullies and stream-cut banks, the thin-layered gray shale can be spotted. Pale yellow **bentonite** bands run through it, layers of a fine clay-like substance formed from volcanic ash modified by later chemical changes. They are a sure indicator that there were active volcanoes somewhere to windward in Cretaceous time, when the shale was being deposited.

To the west, all along the foot of Cheyenne Mountain, the Pierre Shale bumps right into Pikes Peak Granite and other Precambrian rocks, for a major fault separates the Precambrian mountain mass and the Cretaceous sedimentary rocks of the Colorado Piedmont. The granite has pushed eastward here, over some of the sedimentary rocks, for nearly a mile. Paleozoic strata, too weak to stretch in a true anticline across the uplifted and overthrust Precambrian block, are not visible here, although they are present under the Cretaceous layers at a depth of several thousand feet.

The Front Range ends abruptly at the south end of Cheyenne Mountain, although arching sedimentary rocks still reflect the range's uplift for many miles south (see COLORADO 115).

Fountain Creek, east of the highway, flows south toward the Arkansas River, its valley carved in Pierre Shale. Across the valley between Miles 124-108, little conical hills on the eastern skyline are Tepee Buttes, fancifully named mounds of relatively resistant limestone in the Pierre Shale. Up to 50 feet high, the buttes are rich in marine fossils, especially little round clams called *Lucina*. They seem originally to have developed as reef-like mounds on the floor of the

Cretaceous sea, which here in the interior of the continent was never very deep.

About 20 miles south of Colorado Springs, watch to the west for glimpses of the Sangre de Cristo Range 60 miles away behind the Wet Mountains. One of the youngest mountain ranges in Colorado, the Sangres extend in an unbroken rampart from Salida, Colorado to Santa Fe, New Mexico, 235 miles. Their western side is sharply faulted and very steep, with a main fault that is much younger and more active than most others in the state. Indeed, a small fault along the west base has moved within the last few hundred years — just yesterday by the geologic clock.

The large mesa east of Miles 108-102 is called Baculite Mesa, for on its flanks, in the Pierre Shale, are fossils of the straight-shelled Cretaceous ammonite *Baculites*, an ancestor of the modern Chambered Nautilus. This is one case where a geographic feature has been given a paleontologic name; usually it is the other way around. **Paleontology** is the branch of geology devoted to the study of ancient life as shown in **fossils**, which are petrified remains or impressions of ancient animals or plants.

In the last few miles of its southward course, before flowing into the Arkansas River, Fountain Creek cuts deeper and deeper into the Pierre Shale. At the same time, the shale rises (roadcut, Mile 107) as it approaches the south end of the Denver Basin. As you can see on the map, the Pierre hardly occurs south of Pueblo. The highway won't be as bumpy, either, once it leaves the shale roadbed!

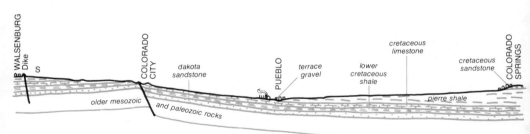

Section along I-25 from Colorado Springs to Walsenburg

As I-25 leaves Pueblo it passes smoky smelters of Colorado Fuel and Iron Works and huge man-made mesas of **slag,** a lava-like waste product of iron processing. The smelters were located here in the 1880s because abundant coal and lime were available nearby in

Cretaceous rocks, water could be obtained from streams draining the Wet Mountains, and iron ore could be brought in easily by railroad from mines in the Sangre de Cristo and Mosquito Ranges. Limestone now is quarried west of Salida but iron ore has been shipped from Wyoming ever since the early supply was depleted.

Between Pueblo and Walsenburg the highway follows the old Taos (New Mexico) Trail across a surface of the Cretaceous limestone that underlies the Pierre Shale. These rocks, too, rise southward, onto a broad gentle upward fold or anticline which crosses this part of Colorado and rims the south end of the Denver Basin. On mesas east of Mile 88 you can see the sloping limestone layers. Gently tipped mesas are more correctly called **cuestas**, a Spanish word for hill. In roadcuts and streambanks you can often get a good look at the impure, light gray limestone interbedded with darker layers of shale. Fossil clams and oysters (*Inoceramus* and *Ostrea*) are common in it. A good place to look for them is in the deep gully between the northbound and southbound Rest Stops at Miles 82 and 81. South of Mile 75 the limestone cuestas slope in the opposite direction, having arched across the crest of the anticline.

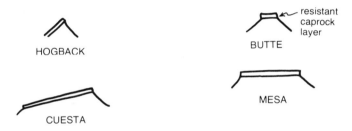

South of Mile 73 the lowest of Colorado's Cretaceous rocks comes to the surface — the Dakota Sandstone, an easily recognized rock made of beach sand spread by the sea as it came westward into Colorado about 100 million years ago. This is the same sandstone that forms the Dakota Hogback, the off-and-on ridge that can be traced along the edge of the mountains from Wyoming to New Mexico. Notice its tawny color and angular way of breaking. It is Colorado's most widespread rock unit, so see if you can learn to recognize it at a glance.

The Wet Mountains, now directly west, are another faulted anticline cored with Precambrian crystalline rocks. On the northeast flank, large blocks of Paleozoic and Mesozoic rocks are faulted against the granite. At the south end of the range, opposite about Mile 63, upturned Paleozoic and Mesozoic rocks describe an

Spanish Peaks are masses of igneous rock intruded through older sedimentary rocks in Tertiary time. They are surrounded by radial dikes, one of which shows clearly in this photograph.
JACK RATHBONE PHOTO

arrowhead-shaped swath around the mountain tip.

A narrow valley southwest of the Wet Mountains is floored with Tertiary sedimentary layers. This valley is particularly interesting to paleontologists studying fossil mammals, for almost perfect skeletons of *Eohippus*, a tiny four-toed ancestor of the horse, have been found there.

The twin domes of Spanish Peaks now jut up to the south. These huge masses of igneous rock pushed and melted their way upward in Tertiary time, possibly never even reaching the surface. They are probably much reduced from their former height now. **Dikes**, thin vertical sheets of igneous rock radiating from them, show that fissures and cracks opened and filled with molten rock material (**magma**) as the peaks formed. The dikes range in width from one to 100 feet, and are up to 14 miles long. Being more resistant to erosion than surrounding sedimentary rocks, they stand as vertical walls above the surface. One of the northernmost dikes crosses I-25 just south of Mile 56.

A particularly prominent dike, one of those radiating from Spanish Peaks, crosses US 85/87 just before that highway enters Walsenburg and the valley of the Cucharas River. It unfortunately is not exposed along I-25.

The lonely cone-shaped hill east of Mile 59 is a small volcanic neck, last remnant of a volcano probably Tertiary in age. Named Huerfano, the Orphan, by some poetic Spanish explorer or settler, it marks the old trail from Denver to Taos, New Mexico. A river, town, and country have inherited its name. JACK RATHBONE PHOTO

I-25
Walsenburg to New Mexico

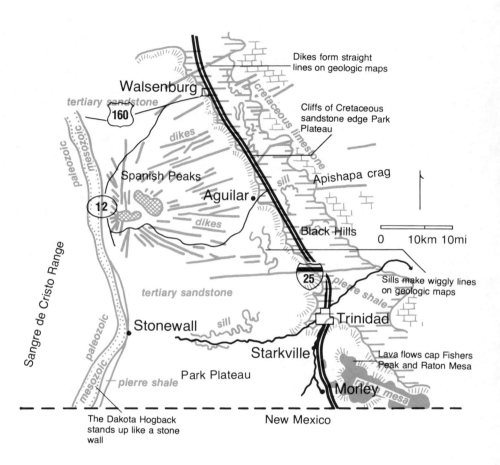

Dikes form straight lines on geologic maps

Cliffs of Cretaceous sandstone edge Park Plateau

Apishapa crag

Black Hills

Sills make wiggly lines on geologic maps

Walsenburg

tertiary sandstone

160

paleozoic

mesozoic

Spanish Peaks

Aguilar

12

dikes

dikes

sill

cretaceous limestone

25

pierre shale

Sangre de Cristo Range

tertiary sandstone

paleozoic

mesozoic

pierre shale

Stonewall

sill

Trinidad

Starkville

Park Plateau

Lava flows cap Fishers Peak and Raton Mesa

Morley

mesa

0   10km 10mi

The Dakota Hogback stands up like a stone wall

New Mexico

interstate 25
walsenburg — new mexico
(57 miles)

Between Walsenburg and Trinidad, Interstate 25 parallels a high, irregular, juniper-covered escarpment of Upper Cretaceous sandstone. All the way to Trinidad the highway is on gray Pierre Shale, which is also Cretaceous but is under and therefore older than the sandstone. Actually the two — the Pierre Shale and the light-colored, blocky sandstone — intertongue. **Intertonguing** or **interfingering** layers are not uncommon geologically. Here they show that the shoreline moved back and forth with changes in sea level. Dark gray marine shales of the Pierre Formation were deposited when the sea covered the area, and beach sands and sands of coastal bars were deposited as it receded. With each new advance of the sea — and probably each advance represents many thousands of years — more marine shales accumulated.

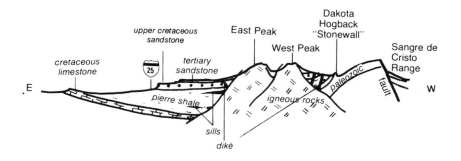

Section across I-25 near Aguilar. Spanish Peaks are enlarged.

As mentioned in the previous section, Spanish Peaks are two great domes of igneous rock that probably pushed their way upward after the faulted anticlines of the Colorado Rockies had formed. Such large intrusions of igneous rock are called **stocks**. The hot, molten rock, under tremendous pressure, elbowed its way through Tertiary sedimentary layers, doming them upward and baking them in a

Near the town of Stonewall the Dakota Sandstone, bent up steeply by the rising Sangre de Cristo Range, forms a nearly vertical rampart. Colorado 12, a scenic route that loops around the Spanish Peaks, passes through a gap in this wall. Dikes form other walls in this region. W.T. LEE PHOTO, COURTESY OF USGS

900-foot-wide zone around the intrusion. A side road west through Aguilar and up and over Cuchara Pass west of the Spanish Peaks makes an interesting side trip, since it crosses part of the Spanish Peaks intrusions and passes over or tunnels through many dikes.

Between Walsenburg and Trinidad the highway crosses some of the many dikes that radiate from Spanish Peaks. The wall-like sheets of rock were formed when molten rock (or **magma**) forced its way into vertical cracks and then cooled. A particularly large dike known locally as Apishapa Crag makes a prominent ridge east of Mile 30. South of it are the Black Hills, composed of Pierre Shale but protected and made resistant by the hard rock of several igneous sills. **Sills** are of the same basic nature as dikes, except that instead of intruding vertical cracks, the magma forced its way between horizontal or nearly horizontal layers of rock. These sills have the same mineral composition as the dikes, and like them are connected to the intrusive masses of Spanish Peaks.

40

Behind the cliffs that border the highway is a broad plateau composed of Cretaceous and Tertiary sandstone and shale deposited in beaches, bars, and coastal swamps during and after retreat of the Cretaceous sea. Wherever there were swamps, there now are coal seams. You will see more of them farther south, for Trinidad owes its existence to coal deposits in Cretaceous and Tertiary sandstone. Coal-bearing rocks extend along the foot of the mountains for many miles, in both Colorado and New Mexico. There are several large active mines west of Trinidad on Colorado 12, and a number of old mining towns south of Trinidad near I-25. Roadcuts along I-25 slice through Cretaceous and Tertiary rocks, revealing the massive fine-grained white sandstones and brownish shales as well as some coal seams. Near Morley, only a ghost town now with ruins of a little church perched above the coal dumps, sandstone and shale, coal seams, and several dikes are clearly exposed in natural exposures, railroad cuts, and particularly large highway cuts. Notice how the shales and sandstones thicken and thin in apparently random manner. Coal seams are quite discontinuous too, as you would expect in a shifting beach-bar-lagoon setting.

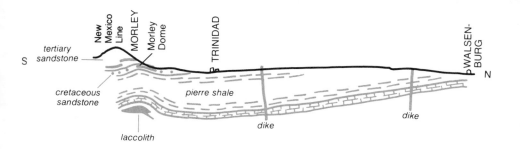

Section along I-25 from Walsenburg to New Mexico line.

Morley, incidentally, is at the center of Morley Dome, and all the strata dip away from this center. The dome is 450 feet high in terms of its geologic structure — geologists would say it has 450 feet of **closure**. It was caused apparently by a small igneous intrusion which pushed its way between rock layers and domed overlying sediments. An oil well drilled on the crest of the dome a few years ago penetrated about 450 feet of igneous rock — a thickness just equal to the closure. An igneous mass of this type, flat-bottomed, round-topped, and doming up overlying beds, is a **laccolith**.

Fisher Peak and Raton Mesa rise dramatically northeast of Morley. On them layers of Tertiary and Cretaceous sedimentary rock, the

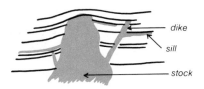

Stocks are large intrusive masses that may or may not dome up overlying rocks.

Laccoliths, smaller than stocks, squeeze between rock layers like sills. They are thick enough to dome up overlying rocks.

same sandstones and shales you have been seeing by the roadside, are protected by a cap of lava. Although you can't see all the lava flows from the highway, there are eleven of them, one on top of another. The lava came from small volcanic vents near La Junta, and remnants of the flows are scattered on mesa tops south and east of here.

Northwestward, Spanish Peaks and the Sangre de Cristo Range, hidden north of Trinidad by the Cretaceous escarpment, come into view. In the Sangres, the long, sharp-toothed line of peaks is lifted by faulting. In this part of the range, Paleozoic sedimentary rocks tilted by the uplift extend clear to the top of the range. Pennsylvanian rocks

Fisher Peak's multi-layered volcanic cap protects underlying Tertiary and Cretaceous sandstone and shale. JACK RATHBONE PHOTO

form many of the summits, and red Permian sandstone 20,000 feet thick forms the upper part of the mountain slope.

Raton Pass, at 8560 feet, has historically been one of the main passes to the south, used by Indian, Spaniard, Frenchman, and Union soldier, by ox-drawn wagon trains and swaying stagecoaches on the Bents Fort branch of the Santa Fe Trail. In 1878 a railroad was completed across it, and a narrow highway built in 1922 can still be picked out among the pinyon and juniper of the surrounding slopes. The pass gets its name from a local inhabitant, the furry-tailed packrat.

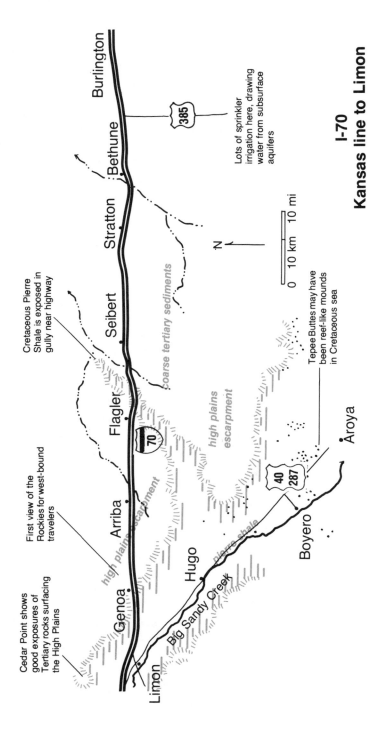

Burlington

Bethune

Lots of sprinkler
irrigation here, drawing
water from subsurface
aquifers

385

Stratton

N

0 10 km
0 10 mi

Seibert

Cretaceous Pierre
Shale is exposed in
gully near highway

coarse tertiary sediments

high plains
escarpment

Flagler

70

Tepee Buttes may have
been reef-like mounds
in Cretaceous sea

Aroya

First view of the
Rockies for west-bound
travelers

high plains escarpment

Arriba

40
287

Boyero

Cedar Point shows
good exposures of
Tertiary rocks surfacing
the High Plains

Genoa

Hugo

Pierre shale

Big Sandy Creek

Limon

I-70
Kansas line to Limon

Center-pivot sprinkler systems in eastern Colorado use water from wells that tap groundwater accumulated above the Pierre Shale. This supply will eventually run out, as it is being used faster than it is recharged.

PHOTO COURTESY BUTLER MFG. CO.

interstate 70
kansas — limon

(80 miles)

Between the Kansas border and a little town called Genoa, Interstate 70 crosses coarse Tertiary gravel that surfaces the central and southern High Plains, gravel carried here by streams flowing eastward from the Rocky Mountains. The High Plains are not nearly as smooth as they seem at a distance; numerous small streams gully them, but unless it has just rained most of these streams are dry. West of the 100th meridian, the approximate dividing line between the moist mid-continent United States and the arid West, agricultural land must be irrigated. Water for the great center-pivot sprinkler systems that circle the fields during the growing season comes from several hundred feet below the surface, where rainwater and snowmelt percolating down through the gravel have accumulated above impermeable Cretaceous shale. The sprinklers use as much as 1000 gallons a minute, and sometimes one well supports two or three systems, so you can see that water is plentiful at that level. However, this is "fossil water" no longer being replaced as fast as it is used, so wells need to be deepened periodically.

45

Genoa lies at the western rim of the High Plains, where the land drops abruptly away into the Colorado Piedmont. In clear weather the mountains are visible from here — the Front Range of the Rockies is 80 miles away. Here the Interstate winds down off the High Plains into the broad valley of Big Sandy Creek.

The High Plains escarpment is quite well defined north of this valley at Cedar Point. A low cliff of Tertiary rocks forms the prow-like point and below it are layers of sandstone laid down as beaches and sand bars edging the Cretaceous sea. Below the sandstone in turn are gray marine shales of the Cretaceous Pierre Shale, deposited when the shallow sea covered the area. Pierre Shale is easy to recognize by its muddy gray color and fine texture, but it converts easily to soil so good exposures are quite rare. Where they do occur, beautiful ammonite shells can sometimes be found, fossilized relatives of our present-day Chambered Nautilus, their mother-of-pearl shell surfaces still shiny and lustrous. Large Cretaceous clams called *Inoceramus*, and bones and teeth from sharks and big swimming reptiles (Mosasaurs and Icthyosaurs) are fairly common as well.

The wide valley of Big Sandy Creek is floored with river floodplain deposits, layered sand and mud and gravel derived from both Tertiary and Cretaceous rocks, deposited whenever the stream overflowed its banks. A few miles downstream (back to the southeast) the valley widens out, and there the Pierre Shale is better exposed and spotted with small, conical Tepee Buttes, actually much flatter than tepees! These buttes may originally have been small patches of reef on the floor of the shallow Cretaceous sea.

interstate 70
limon — denver
(89 miles)

Between Limon and Denver, Interstate 70 crosses the Colorado Piedmont, traversing younger and younger rocks that fill the center of the Denver Basin. Just west of Limon, as the highway swings northwest toward the town of River Bend, it leaves the Big Sandy and rises a little bit to cross the drainage divide between Arkansas River drainage (Big Sandy Creek) and drainage of the South Platte River (Bijou Creek). The South Platte itself is about 65 miles north of here. Most of its tributaries flow north, and most are sporadic intermittent streams that run only in spring or after heavy rains.

As the highway rises across the drainage divide it leaves the gray Cretaceous Pierre Shale and passes onto higher, younger sandstone layers that were once beaches and sandbars along the edge of the retreating Cretaceous sea. Cedar Point, a prow-like prong of the High Plains, rises to the north. Cretaceous sandstone forms poor soil, used mostly for grazing land. Some of the sandstone contains thin layers of coal, evidence of marshes and lagoons along the Cretaceous shore.

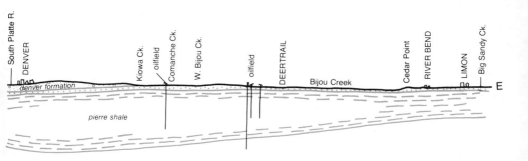

Section along I-70 between Limon and Denver.

Between Deer Trail and Byers the Interstate crosses to rocks of the Dawson Formation, a series of sandstone and conglomerate layers that span the end of the Cretaceous Period and the beginning of the

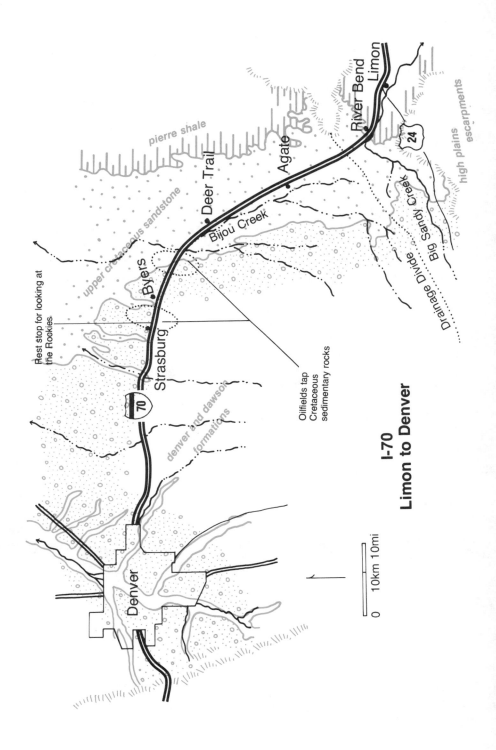

pierre shale

Deer Trail

Agate

River Bend

Limon

high plains escarpments

Big Sandy Creek

Drainage Divide

24

upper cretaceous sandstone

Bijou Creek

Byers

Rest stop for looking at
the Rockies

Strasburg

70

denver and dawson
formations

Oilfields tap
Cretaceous
sedimentary rocks

Denver

I-70
Limon to Denver

0   10km 10mi

48

Tertiary. When these rocks were deposited, the Denver Basin, bounded on the west by the rising Front Range, was already well defined; as it sank, the sandstone and conglomerate filled its center. Between Miles 320 and 310, a few wells penetrate the Denver Formation to tap oil accumulated in Cretaceous shale and sandstone beneath it. More productive Denver Basin oil fields lie farther north.

Much of the irrigation water in the Piedmont area comes from shallow wells tapping layers of loose Pleistocene gravel that in some places form a veneer over older rocks. The gravels accumulated during glacial episodes in the mountains, when rivers were greatly over-burdened with sand and rock dumped into them by melting ice.

Water rights here are a major legal issue. The rights to surface water — water in streams and rivers — were established by law soon after the area was settled. Much later, when wells were drilled, farmers learned that pumping from wells often depleted stream flow. Such indirect interference with prior water rights has caused some knotty problems for Colorado courts, which often call on geologists to serve as "expert witnesses" in cases like these.

By now you should be able to see the Rockies clearly unless of course they are cloud-covered, as they may well be in winter or on summer afternoons. To pioneers coming from the east this rampart was the front of the mountains, so they called it the Front Range. To the northwest is Rocky Mountain National Park, an especially scenic area crossed by the highest through highway in United States (see US 34). Longs Peak, dead ahead at Mile 288, is 14,255 feet in elevation, the highest point in the park.

Directly west, beyond Denver, the mountains are more jagged, without the high rolling uplands characteristic of Rocky Mountain Park. The highest peak there is another Fourteener, Mt. Evans, 14,264 feet above sea level, with Mt. Bierstadt close behind it. The road up Mt. Evans is considered to be the highest automobile road in the country (see DENVER AREA). Seventy-five miles away to the southwest you may be able to see the rounded 14,109-foot summit of Pikes Peak (see COLORADO SPRINGS AREA). Ancient Precambrian rocks, hard and crystalline, form all of these peaks as well as the main mountain mass.

You can also see the Tertiary pediment from here, an irregular horizontal surface that forms a line about halfway up the mountains, at about 9000 feet elevation. Don't confuse it with timberline, the top of the dark evergreen forest, which is closer to 11,000 feet. In winter the snowline may deceive you too.

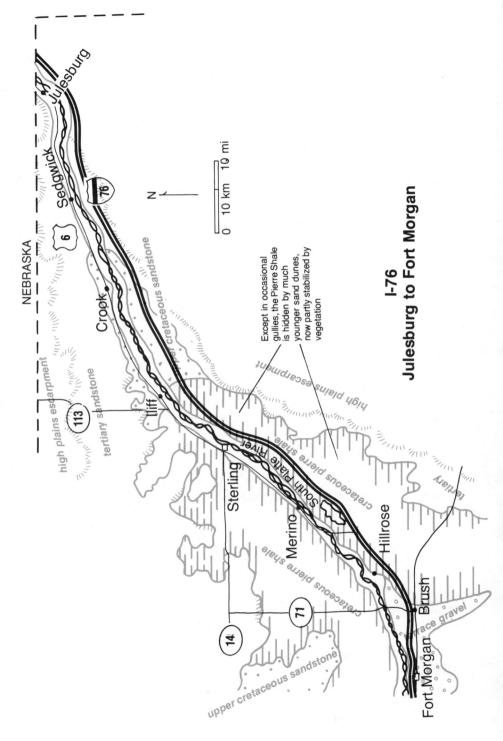

**I-76
Julesburg to Fort Morgan**

Except in occasional gullies, the Pierre Shale is hidden by much younger sand dunes, now partly stabilized by vegetation

NEBRASKA

Julesburg

Sedgwick

Crook

Iliff

Sterling

Merino

Hillrose

Brush

Fort Morgan

high plains escarpment

tertiary sandstone

upper cretaceous sandstone

cretaceous pierre shale

high plains escarpment

South Platte River

cretaceous pierre shale

tertiary

upper cretaceous sandstone

terrace gravel

N

0   10 km   10 mi

50

interstate 76
julesburg — fort morgan
(111 miles)

Interstate 76 enters Colorado on the floodplain of the South Platte River, and follows the route of the old Overland Trail and stage route established in the mid-1800s. Low hills on either side are surfaced with Tertiary sedimentary rocks, loosely consolidated sand and gravel that surface the High Plains. Soon after entering the corner of the state the highway rises onto these hills for a time (Mile 164) and then passes almost imperceptibly (Mile 134) into a very broad, shallow valley of older rocks, the Colorado Piedmont. To the north, the tan bluffs of the High Plains escarpment can occasionally be glimpsed; they gradually become more and more distant from the highway, for they follow a roughly east-west line parallel to the Nebraska boundary.

Between Crook and Hillrose the highway crosses an eroded surface of Cretaceous rocks, first Upper Cretaceous sandstone that formed on beaches and sand bars of the receding sea, then dark gray Pierre Shale, often containing fossil shells of marine animals that lived on or sank to the muddy sea floor. Outcrops of these rocks are rare because this entire area is covered with sand hills, one-time dunes now stabilized by vegetation (and in some recent "blowouts" by old automobile tires). The dunes are thought to have developed during the Ice Ages when glacier-fed streams overflowing their banks deposited floodplains of sand and gravel, and strong winds sweeping down from mountain icefields carried the sand from the riverbanks and piled it into dunes. Some sand must have come from loosely cemented Upper Cretaceous sandstone as well. Patches of sand hills extend most of the way to Denver and you may have noticed that they cover large parts of western Nebraska too.

Just southwest of Hillrose (Mile 92) the highway descends from the sand hills onto a broad river terrace that is 60 to 80 feet above the level of the South Platte River. This terrace, the site of the towns of Brush and Fort Morgan, is composed of river-deposited gravel and sand. It is an old river floodplain formed at about the same time as the dunes, during a long period when the river was heavily charged with

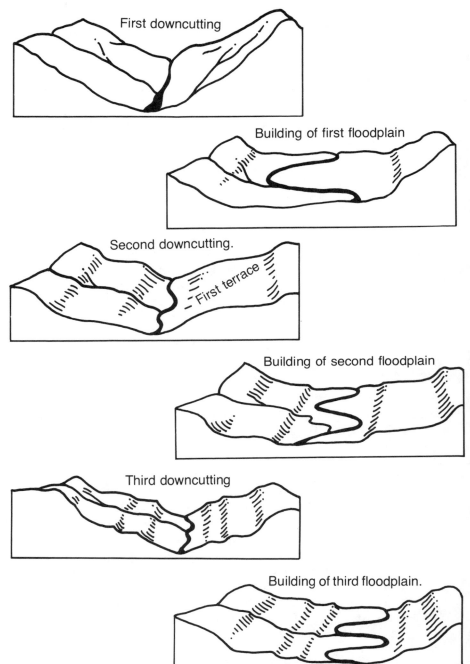

First downcutting

Building of first floodplain

Second downcutting.

First terrace

Building of second floodplain

Third downcutting

Building of third floodplain.

Second terrace has been eroded away on one side

As many as three terrace levels border some streams in Colorado. Each represents a fairly stable period during which an overburdened stream widened its channel and filled it partway with rock debris.

rock debris and habitually overflowed its banks. In places several terrace levels occur, each representing a similar period of stabilized but none-the-less heavy river flow. These alluvial terraces can be correlated with others nearer to the mountains, and those in turn can sometimes be traced continuously along major streams into the mountains, where each terrace seems to end at a moraine left by an ancient glacier. So we can be fairly sure that terrace formation coincides with periods of cooler (perhaps almost arctic) climate, when there were glaciers in mountain valleys, torrential heavily over-loaded streams often reaching flood stage, and cold winds sweeping down from the mountains, picking up sand and piling it into sand dunes.

An interesting sidelight: West of here, close to the mountains, a famous archeological locality known as the Lindenmeier Site is on one of these terraces close up under the High Plains Escarpment. Human artifacts found there suggest that, sometime between 12 and 13 thousand years ago, early Americans used the terraces as sites for temporary camps.

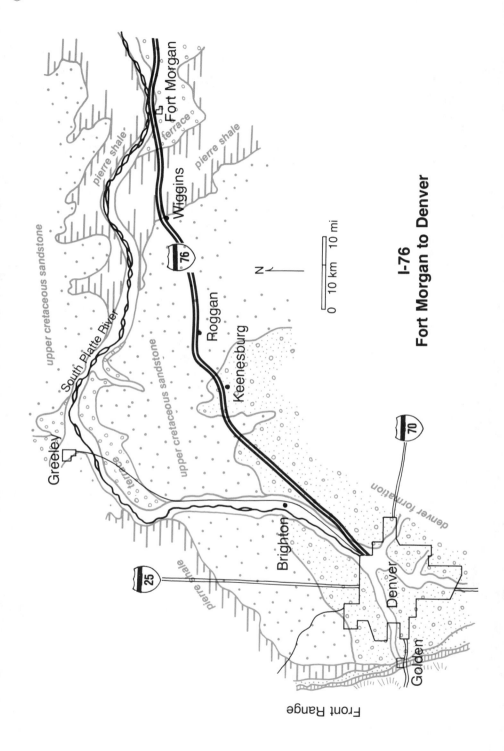

I-76
Fort Morgan to Denver

Fort Morgan

Wiggins

76

Roggan

Keenesburg

70

Greeley

South Platte River

upper cretaceous sandstone

upper cretaceous sandstone

pierre shale

pierre shale

terrace

Brighton

Denver

denver formation

25

pierre shale

Golden

Front Range

N

0   10 km   10 mi

54

interstate 76
fort morgan — denver
(84 miles)

Leaving Fort Morgan, Interstate 76 travels for a time on the South Platte River terrace and then leaves the river and climbs slightly into monotonous sand-hill country. For half of the distance between Fort Morgan and Denver, the sand hills cover Cretaceous sedimentary rocks deposited in a shallow sea and on the beaches and bars that bordered it as it receded eastward at the end of the Cretaceous Period. The Rocky Mountains had not yet begun to rise when they were deposited; they thicken westward in the direction of the already rising Wasatch Range of Utah. They are less than 3000 feet thick near the Nebraska line (where they have been sampled by oil well drilling), 7000 feet thick in the deepest part of the Denver Basin, and 20,000 feet thick in North Park on the west side of the Front Range.

There are a few clusters of oil wells along the highway, although I-76 manages to slip between the two areas of greatest production, south of Fort Morgan and north of Sterling. Oil and gas come from the Cretaceous rocks.

After about Mile 59, you will have continuing good views of the Front Range of the Colorado Rockies, depending of course on the weather and the time of day. Due west, the tall prominent peak is Longs Peak, elevation 14,255 feet. North of it are summits of the Mummy Range, usually snow-capped in spring and early summer and often with snow lasting year-round. Longs Peak and the Mummy Range are both in Rocky Mountain National Park (see U.S. 34). South of Longs Peak rises the jagged skyline ridge of Indian Peaks. The prominent broadly pyramidal peak west of Denver, just left of the highway's direction near Mile 51, is Mt. Evans, 14,264 feet above sea level (see DENVER AREA). And far away to the south, seeming to stand east of the general mountain trend, is the rounded summit of Pikes Peak, 14,109 feet in elevation (see COLORADO SPRINGS AREA). All these peaks are composed of the ancient Precambrian rocks that make up the core of the Front Range.

55

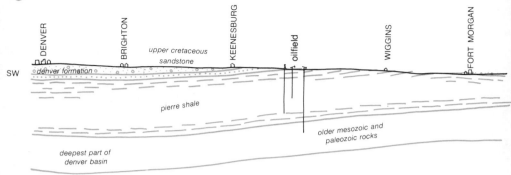

Section along I-76 from Fort Morgan to Denver.

At Mile 44 the Interstate passes out of Cretaceous sedimentary rocks and onto the Denver Formation. On it, tilled soils lose their sandiness and become clayey, but the rocks themselves are still concealed by soil or sand hills. The Denver Formation is on top of other Cretaceous sediments, and is younger than they, containing both Cretaceous and Tertiary rocks. However, it is older than the rocks that surface the High Plains to the east and north.

Looking again at the mountains see if you can distinguish the Tertiary pre-uplift pediment. It is at about 8000 to 9000 feet altitude, halfway up the mountain slope. Of course it is now much altered by erosion, but its hilly surface is horizontal when seen from a distance and forms a fairly level step on the front of the mountains. It is particularly evident northwest of Boulder and west of Denver.

u.s. 24
limon — colorado springs
(71 miles)

Between Limon and Colorado Springs, U.S. 24 crosses from the High Plains to the Colorado Piedmont, paralleling the drainage divide that separates South Platte River tributaries from those of the Arkansas River. The highway climbs out of the valley of Big Sandy Creek, a tributary of the Arkansas, onto a corner of a huge isolated mesa capped with Quaternary gravel.

One look at the wide valley of Big Sandy Creek, with its diminutive trickle of water, will raise doubts about things geological! How could such a stream, usually dry or nearly dry, scour out a valley eight to fifteen miles wide and nearly 100 feet deep, even through loosely consolidated Tertiary and Cretaceous sediments? Much of the scouring may have been accomplished during the Pleistocene Epoch — Ice Age time — when rainfall, snowfall, and runoff far exceeded that of today. But even that does not seem to adequately explain the size of the valley, especially since the Big Sandy does not drain a glaciated area.

Southeast of Limon, from the top of the High Plains island, there is a good view northwestward toward Cedar Point, a western prong of the High Plains. There you can see distinctly the Tertiary rocks which form the High Plains surface, and the sharp, cliff-edged High Plains Escarpment.

For six miles U.S. 24 crosses the gravelly mesa surface; then it descends to Matheson. Altitude-wise, this part of the Piedmont is pretty high: Simla, almost on the South Platte-Arkansas drainage divide, is over 6100 feet in altitude.

The Piedmont here is surfaced with Dawson Formation, thick layers of Tertiary sandstone that fill the center of the Denver Basin. Most sandstone is made primarily of quartz grains, but this sandstone also contains lots of grains of feldspar, suggesting that it was derived from granite, a rock containing both feldspar and quartz grains. Since feldspar readily breaks down into clay, its presence here indicates that the sand didn't travel very far from its original source

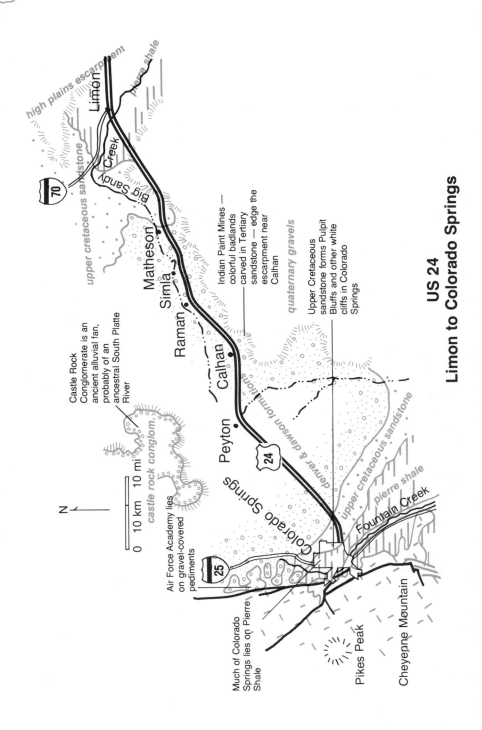

US 24
Limon to Colorado Springs

58

before being deposited. This you might easily guess with the mountains so close at hand.

Looking northwestward notice the flat-topped, forested mesas and buttes on the skyline. They are capped with a resistant, very coarse rock, the Castle Rock Conglomerate, also derived from the nearby mountains. This conglomerate occurs only in a limited area not far from the mouth of West Monument Creek, where the ancestral South Platte River probably emerged from the Rampart Range, now visible to the west. It may be part of an alluvial fan deposited by the ancestral South Platte before the river changed course to flow northward toward Denver.

Between Matheson and Calhan the highway skirts a steep escarpment, which at places is only a mile or two southeast of the road. At Calhan, erosion of this escarpment has bared several little badland areas with gaily colored ravines and gullies and standing rock pillars or **hoodoos**, an area known locally as "Indian Paint Mines." There is evidence that Indians frequented this area — many arrowheads and spearpoints have been found — but there is no direct evidence that they used the colored clays for war paint or anything like that. Most of the yellow, red, and purple rock is pigmented with iron oxides; gypsum crystals whiten some layers. Around 1888 clay was dug here for firebricks, and some of it has supposedly been used for pottery.

The ever-closer views of Pikes Peak from this highway are truly breathtaking. In clear weather every cliff and crag and slope stands out sharply. Two miles west of Calhan, the view to the southwest also includes the sawtooth crest of the Sangre de Cristo Range 100 miles away. The country here opens out into typical Colorado Piedmont — rolling, farm-strewn hills divided by intermittent streams.

Pikes Peak, lone sentinel of the south end of the Front Range, is 14,109 feet high. Although not the tallest mountain in Colorado, it rises nearly a mile and a half above its base, more than any other peak in the state. The Pikes Peak **massif** (a word used by geologists to indicate the whole of a mountain mass which behaves geologically as a single unit) includes Cheyenne Mountain, the flat-topped smaller mountain to the south. Rampart Range, stretching in a long low horizontal line northward, is composed of the same type of granite as Pikes Peak, but is separated from the massif by a giant northwest-southeast break, the Ute Pass Fault. The flat, almost horizontal tops of Cheyenne Mountain and Rampart Range are parts of the old Tertiary pediment formed before the last uplift of this region.

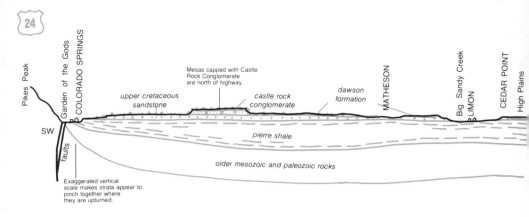

Pikes Peak

Garden of the Gods

COLORADO SPRINGS

upper cretaceous sandstone

Mesas capped with Castle Rock Conglomerate are north of highway.

castle rock conglomerate

dawson formation

MATHESON

Big Sandy Creek

LIMON

CEDAR POINT

High Plains

SW

Faults

pierre shale

older mesozoic and paleozoic rocks

Exaggerated vertical scale makes strata appear to pinch together where they are upturned.

Section along and north of US 24, Limon to Colorado Springs

As the highway approaches Colorado Springs it drops into sandstone laid down on beaches, bars, and coastal floodplains as the Cretaceous sea retreated from this area. Below the sandstone is gray fossil-bearing marine shale of the Pierre Formation, deposited over a vast area before the sea began to withdraw. Lower parts of the city are built on this shale; higher parts lie on the sandstone or on terraces of sandstone topped with younger gravels. The terrace-like hills close to the mountains are true pediments, levels carved into the original bedrock of the mountain. They also are topped with gravel.

The Tertiary pediment shows up well on Cheyenne Mountain, just southeast of Pikes Peak, as shown in this aerial photograph.

T.S. LOVERING PHOTO, COURTESY OF USGS

60

u.s. 50

kansas — la junta

(88 miles)

U.S. 50 enters Colorado on the floodplain of the Arkansas River near Colorado's lowest point, 3300 feet above sea level. A fairly sizable river for this part of the country, the willow- and cottonwood-bordered Arkansas drains much of southeastern Colorado as well as the area between the Front Range and ranges farther west. For most of the distance to La Junta the highway remains in sight of the river, following the one-time route of the Santa Fe Trail across sand and gravel of the river floodplain or dark gray Cretaceous shale bordering the floodplain. Low hills north of the river are surfaced with Cretaceous limestone, a little younger than the shale but hardly differing in appearance from the shale slopes that extend to the south. Thin beds of light-colored limestone are visible in some roadcuts.

Most of the tributary streams are intermittent: they flow only seasonally or after heavy rains. The Arkansas itself does not flow heavily any more because much of its water is drawn off in irrigation ditches and canals. Reservoirs store abundant spring runoff for summer use. The irrigated floodplain produces sugar beets, soy beans, corn, alfalfa, and garden crops, whereas the non-irrigated slopes north and south of the floodplain are used for grazing. Sand-hills and small dune areas edge these slopes. A hundred years ago huge herds of cattle driven north from Texas over a fenceless land crossed the Arkansas River here.

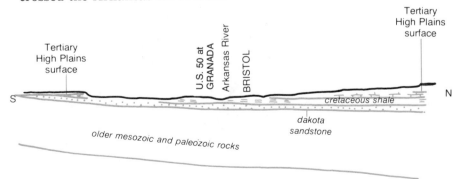

Section across U.S. 50 at Granada

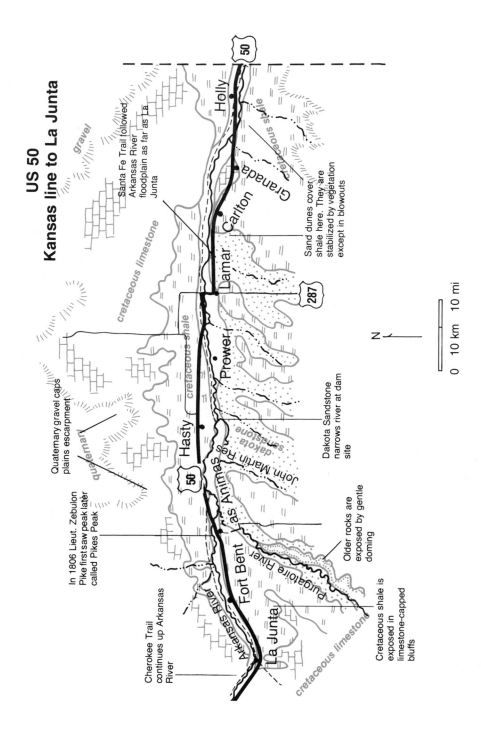

US 50
Kansas line to La Junta

gravel

Santa Fe Trail followed Arkansas River floodplain as far as La Junta

cretaceous limestone

Holly

Granada

Carlton

Lamar

cretaceous shale

Sand dunes cover shale here. They are stabilized by vegetation except in blowouts

287

Prower

dakota sandstone

cretaceous shale

Dakota Sandstone narrows river at dam site

N

0   10 km   10 mi

Quaternary gravel caps plains escarpment

quaternary

Hasty

50

John Martin Res.

In 1806 Lieut. Zebulon Pike first saw peak later called Pikes Peak

Las Animas River

Fort Bent

Purgatoire River

Older rocks are exposed by gentle doming

Cherokee Trail continues up Arkansas River

Arkansas River

La Junta

cretaceous limestone

Cretaceous shale is exposed in limestone-capped bluffs

50

Alternating layers of limestone and shale seem to represent rapid and repeated changes from muddy water, in which the dark clay was deposited, to clear water, in which the limestone was laid down. Similar changes characterize a region that extends more than 2000 miles in a north-south direction. Note figure for scale.
P.T. VOEGELI PHOTO, COURTESY USGS

In river bluffs below John Martin Reservoir Dam the Dakota Sandstone, Colorado's oldest widespread Cretaceous rock, is well exposed. (Turn south to the dam at Hasty.) Older than the gray shale, this sandstone is more resistant than surrounding sedimentary rocks, so it tends to form cliffs and narrow canyons. It is responsible here for narrowing the Arkansas River channel and forming a good dam site. As can be seen along the road to the recreation area below the dam, the sandstone is cross-bedded, marked by diagonal laminae that record currents that sorted the sand during deposition.

All the rock units except the gravels that top the terraces visible in the distance to the north dip very gently northward into the great sag of the Denver Basin, showing us that the basin formed in late Cretaceous and early Tertiary time.

One of the area's few permanent streams, the Purgatoire River, enters the Arkansas near Las Animas. Fed from the Sangre de Cristo Mountains to the west and the high country around Mesa de Maya to the south, it drains a large and unusually colorful region. With its tributaries it cuts across a gentle dome of sedimentary rocks ranging in age from Late Paleozoic to Cretaceous. Miniature Grand Canyons form a twisting maze of pink sandstone cliffs and dark red shale slopes. However, few roads lead into the deeply incised wonderland.

West of Las Animas the slopes both north and south of the river are composed of dark gray Cretaceous shale capped with light gray Cretaceous limestone. Here, rocks dip westward. Eventually the shale dips below the surface of the river floodplain, and cliffs and ledges of overlying limestone converge on the highway. The limestone actually consists of alternating thin layers of limestone and dark shale, as many as 30 alternations within 20 feet. Geologists feel that they must result from some very widespread changing conditions, perhaps rapid fluctuations in sea level or climate.

To the southwest rise the two Spanish Peaks, great igneous intrusions formed soon after the lifting of the rest of the Rockies. These peaks are particularly interesting in that they are surrounded by an unusual radial array of dikes, vertical sheets of igneous material that solidified in cracks in the rock around the intrusions. Being composed of hard, erosion-resistant igneous rock, the dikes now often stand up as walls 10 to 100 feet high.

u.s. 50
la junta — canyon city
(103 miles)

Highway U.S. 50 remains close to the Arkansas River all the way to Pueblo. Even west of Pueblo it is never more than a few miles from the river.

Near La Junta, light-colored Cretaceous limestone interbedded with thin layers of shale (see previous section) covers the slopes both north and south of the river. The limestone accumulated in a broad sea that covered Colorado 80 to 90 million years ago. It dips gently northward into the Denver Basin. On the geologic map the curving contact between the Pierre Shale and the underlying limestone neatly outlines the south end of the Denver Basin. West of Pueblo the rocks rise onto a gentle upward fold or anticline that closes off the southwest edge of the basin, and continues northward into the huge mountain-forming faulted anticline of the Front Range. In a geologic sense, this inconspicuous fold is the southern tip of the Front Range.

Many fossils of marine animals occur in the Cretaceous limestone. The most common are tiny bead-like shells of one-celled foraminifera. They occur with much larger ammonites, which are relatives of the modern Chambered Nautilus and octopus, and with *Inoceramus,* a large clam with a corrugated shell.

Shortly before you get to Rocky Ford you may get your first glimpses of Pikes Peak just to the right of the highway. In a precursor to this volume, N.H. Darton, writing in 1916 about the geology along the Santa Fe Raiload route, speaks of Rocky Ford's famous cantaloupes. Now, lettuce, onions, sugar beets, corn, fruit, and other garden crops are grown here also.

About 12 miles west of Rocky Ford, younger rocks come to the surface: the Upper Cretaceous Pierre Shale, a gray fine-grained rock deposited as sea-bottom mud. The shale is hardly ever well exposed, but it is very thick, 4000 feet just north of here and thickening westward.

65

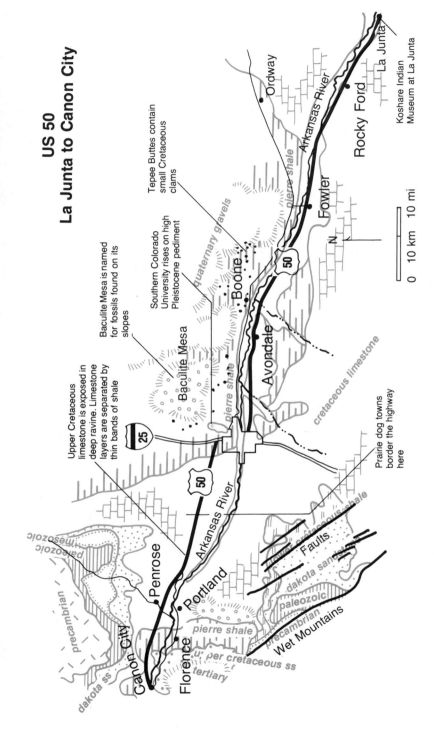

US 50
La Junta to Canon City

Tepee Buttes contain small Cretaceous clams

Baculite Mesa is named for fossils found on its slopes

Southern Colorado University rises on high Pleistocene pediment

Upper Cretaceous limestone is exposed in deep ravine. Limestone layers are separated by thin bands of shale

Prairie dog towns border the highway here

Koshare Indian Museum at La Junta

Ordway

Arkansas River

Rocky Ford

Fowler

La Junta

pierre shale

quaternary gravels

Boone

Avondale

cretaceous limestone

Baculite Mesa

pierre shale

25

50

50

Arkansas River

Penrose

Portland

Florence

Canon City

paleozoic

mesozoic

precambrian

dakota ss

pierre shale

upper cretaceous ss

tertiary

dakota sandstone

paleozoic

precambrian

Wet Mountains

cretaceous shale

Faults

N

0   10 km   10 mi

Congregations of marine shellfish, possibly living in patches of sea grass, built reef-like aggregates on the floor of the shallow Cretaceous sea. Today these limy aggregates, more resistant than the surrounding shale, form Tepee Buttes.   G.R. SCOTT PHOTO, COURTESY OF USGS

East of Avondale, little cones of harder material occasionally rise above the shale surface. These cones, called Tepee Buttes, resist erosion because they contain more limestone than the surrounding soft shale. They are good places to look for fossils, particularly small clams named *Lucina*. Hundreds of the buttes are scattered north of the river near Avondale, just east of Pueblo, and north and east of Canon City.

Avondale perches 30-40 feet above the river on an Ice-Age terrace composed of gravel and sand and small boulders washed from the Front Range when the Arkansas River was overloaded with debris from glacier-fed tributaries. The terraces are fragments of old flood-plains and represent fairly stable periods in the development of the river, when downcutting was minimal and the river swung back and forth, widening its channel, flooding frequently and spreading flood deposits on either side. How many terrace levels are there here?

Southern Colorado University, at Pueblo, stands on a similar but higher pediment. Northeast of it a still higher mesa with gray slopes is Baculite Mesa, so-named because on its slopes, in the gray Pierre Shale, there are fossil *Baculites,* straight-shelled Cretaceous relatives of the octopus and the Chambered Nautilus. Coarse Pleistocene gravel caps the mesa.

Near Pueblo the highway climbs up out of the Arkansas Valley Pierre Shale onto Cretaceous limestone — the same fossil-bearing white limestone, mixed with shale, that you saw at La Junta and Rocky Ford. Geologically, the road goes **down** into the limestone, since it really underlies the Pierre Shale. But here the limestone is curving over the south end of the Front Range anticline, so it slopes up toward the mountains, topping many gently sloping cuestas. The best exposures of the limestone are in deep ravines between the cuestas. Across the Arkansas River south of Penrose, near the town of Portland (named after Portland, England, original home of Portland cement), Cretaceous limestone is quarried, crushed, and made into cement. Gypsum, another component of cement, is obtained from Pennsylvanian rocks near Coaldale; here it is used to make plaster and plasterboard as well.

Northward, Pikes Peak dominates the landscape. Both this peak and its lower neighbor Cheyenne Mountain are parts of the same huge mass of Pikes Peak Granite, an intrusive igneous rock that formed part of the Precambrian ranges in this area about a billion years ago. The flat top of Cheyenne Mountain, as well as the broad shelf extending westward from Pikes Peak, are parts of the Tertiary pediment, an erosion surface once continuous with the High Plains. The Wet Mountains to the south have a similar nearly horizontal Tertiary surface.

At the south end of the Front Range all the rocks you have been seeing, as well as older ones so far hidden from view, are caught up in the structure of the Front Range, and bend up dramatically at the mountain's edge. The up-ended rock layers form hogbacks that curl around the end of the Precambrian mountain mass. We can imagine that they once swooped on up as a titanic anticline over the top of the

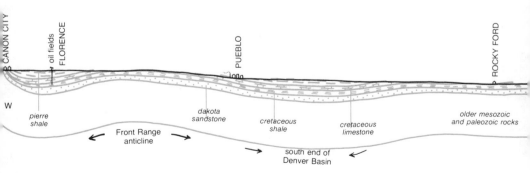

Section along U.S. 50 from La Junta to Canon City

mountains, but in fact they probably eroded away nearly as fast as the mountains rose. You can see the tipped-up rocks, particularly the erosion-resistant layers of Dakota Sandstone, north of the highway. A short side trip up Colorado 67 toward Phantom Canyon will take you right through the hogbacks for a closer view.

Directly north of Canon City the hogbacks formed by these sedimentary rocks bend sharply and run south past Canon City, where the most prominent of them, the Dakota Sandstone, forms the west wall of the penitentiary grounds and the ridge traversed by Skyline Drive. They then continue south along the northeast flank of the Gore Hills and Wet Mountains. Except for some stretches where Cretaceous rocks are absent because of faulting, the Dakota Hogback can be traced clear to the New Mexico line. North of Canon City, many fossil dinosaur skeletons have been dug out of the Morrison Formation, a series of Jurassic shales and sandstones between the Dakota Hogback and the mountains.

Gray shales seen in roadcuts and on mesa slopes east of Canon City are Pierre Shale. Often beveled across their dipping bedding planes, they are Quaternary pediments covered with Pleistocene gravel.

Zebulon Pike, after whom Pikes Peak was named, explored this area for the U.S. government in December 1806 and January 1807, before continuing up the Arkansas River toward Salida.

iii
stairway to the sky
— denver and
colorado springs areas

The cities and towns along the east side of the Rockies are well placed bases for looking at the geology of the Front Range. Many lie close to the great faults that edge the range. Mountains dominate the view to the west, rising with startling suddenness from the mile-high Piedmont to the Tertiary pediment at 8000 to 10,000 feet, and then to summits 12,000 to 14,000 feet above sea level. Far east of the towns, the edge of the High Plains swings in an embracing arc from Wyoming to Colorado Springs.

From Denver and Colorado Springs short trips take you to see the geologic features of this plain-meets-mountain area. Trips into the Front Range show the make-up of a typical faulted anticline range, complete with a core of very old Precambrian crystalline rocks — igneous and metamorphic granite, gneiss, and schist — outlined by rows of upturned sedimentary rocks — shale, sandstone, limestone, and conglomerate. The Tertiary pediment and valleys carved by Ice-Age glaciers vie for attention with rough-and-ready mining history and evidences of recent tragic flash floods. Fossils, records of ancient life, can be found within easy driving distance, as can museum quality mineral specimens.

Short geologic sketches of Red Rocks Park and Mt. Evans, and a longer loop to Boulder, Estes Park, the Peak-to-Peak Highway, and Central City, each show facets of the geologic story of the Denver area. For the Colorado Springs region there are short trips to Garden of the Gods, Williams Canyon and Cave of the Winds, and Pikes Peak, and a longer one to Cripple Creek and Gold Camp Road. Arrange to drive up Pikes Peak or Mt. Evans in the morning, for clouds often close in by afternoon.

For a general picture, a skeleton view on which to superimpose geologic details, read or reread "This Boundless Land — Plains and Piedmont" and "Purple Mountain's Majesty — Folded and Faulted Ranges." The following itineraries in these chapters also describe Front Range geology:

I-25 Wyoming — Denver
I-25 Denver — Colorado Springs
I-70 Denver — Dillon
U.S. 24 Colorado Springs — Buena Vista
Florissant Fossil Beds National Monument
U.S. 24 Limon — Colorado Springs
U.S. 34 Loveland — Granby via Rocky Mountain
   National Park
U.S. 285 Denver — Fairplay
U.S. 287 Wyoming — Denver
Co. 115 Colorado Springs — U.S. 50

The Denver Museum of Natural History, the Geology Museum at the Colorado School of Mines in Golden, and Henderson Museum at the University of Colorado in Boulder have geologic exhibits worth visiting.

the denver area

More than 12,000 feet of sedimentary rocks underlie Denver. Layers and layers of limestone, shale, and sandstone, they are covered with a thin veneer of loose gravel and soil. They do come to the surface, though, close to the mountains, rising up steeply as if they were going to arch across the Precambrian core of the Front Range. West of Denver, the eroded edges of these tilted rocks make rows of steep hogbacks alternating with low, smooth swales.

The lowest layers of rocks, Paleozoic and Mesozoic in age, bend down into the Denver Basin, which is deepest very near the city. The center of the Denver Basin is filled to overflowing with Tertiary sand and gravel, most of it also tilted and reflecting the sagging of the basin. The uppermost layer is the Denver Formation, which filled the very center of the basin and overflowed its edges after the downward sagging had stopped. Its upper surface is now eroded by the South Platte River and its tributaries, which have sculptured the hilly Piedmont surface on which Denver sprawls.

In 1965 the South Platte was swept by a sudden flood that amply demonstrated its abilities as an agent of erosion. The flash flood destroyed or damaged buildings, bridges, machinery, cars, and houses on the low, flat river floodplain, caused $508,000,000 in damage, and drowned six people. In spite of the ever-present danger of a repeat performance, the floodplain has been rebuilt and is again the site of busy industrial activity. The following description of the flood, by H.F. Matthai of the U.S. Geological Survey, stands as a warning:

> The deluge began, not only near Dawson Butte (30 miles south of Denver), but also at Raspberry Mountain, 6 miles to the south, near Larkspur. The rain came down harder than any rain the local residents had ever seen, and the temperature dropped rapidly until it was cold. The quiet was shattered by the terrible roar of the wind, rain, and rushing water. Then the thudding of huge boulders, the snapping and tearing of trees, and the grinding of cobbles and gravel increased the tumult. The small natural channels on the steep slopes could not carry the runoff, so the water took shortcuts, following the line of least resistance. Creeks overflowed, roads became rivers, and fields became lakes — all in a matter of minutes.

The flow from glutted ravines and from fields and hillsides soon reached East and West Plum Creeks. The combined flow in these creeks has been described as awesome, fantastic, and unbelievable; yet none of these superlatives seems adequate to describe what actually occurred. Large waves, high velocities, crosscurrents, and eddies swept away trees, houses, bridges, automobiles, heavy construction equipment, and livestock. All sorts of debris and large volumes of sand and gravel were torn from the banks and beds of the streams and were dumped, caught, plastered, or buried along the channel and flood plains downstream. A local resident stated, 'The banks of the creek disappeared as if the land was made of sugar.'

The flood reached the South Platte River and the urban areas of Littleton, Englewood, and Denver about 8 p.m. Here the rampaging waters picked up house trailers, large butane storage tanks, lumber, and other flotsam and smashed them against bridges and structures near the river. Many of the partly plugged bridges could not withstand the added pressure and washed out. Other bridges held, but they forced water over approach fills, causing extensive erosion. The flood plains carried and stored much of the flood water, which inundated many homes, businesses, industries, railroad yards, highways, and streets.

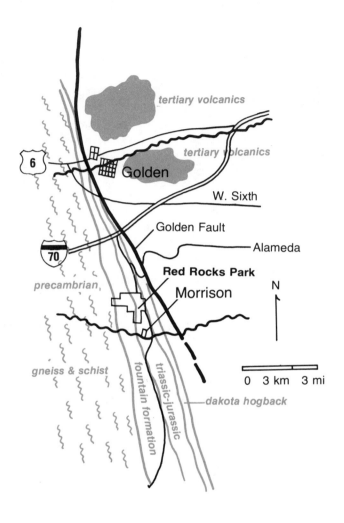

Red Rocks Park and surrounding area

The dramatic red monoliths at Red Rocks Park are composed of layered sandstone, shale, and conglomerate of the 300-million-year-old Fountain Formation. Originally formed on the flanks of the Ancestral Rockies, the layers were tilted up along the mountain flank 65-45 million years ago when the present Rockies were formed. T.S. LOVERING PHOTO, COURTESY OF USGS

red rocks park

(16 miles from State Capitol, Denver)

At Red Rocks Park west of Denver, near Morrison, Pennsylvanian and Permian sedimentary rocks of the Fountain Formation are eroded into spectacular red monoliths, some of them now framing Red Rocks Amphitheatre and providing an unusual dramatic backdrop for summer concerts and operas. The rock of the monoliths consists of layer upon layer of coarse brick-red sandstone and conglomerate with many thinner layers of dark red siltstone between. Festooned with diagonal striations called **cross bedding**, at an angle to the main layers (or **beds**) of the rock, the sandstone is of a type originally deposited by torrential streams. The cross bedding and other evidence show that the rock was part of an apron of debris washed off the east edge of the Ancestral Rocky Mountains nearly 300 million years ago.

The great red monoliths were carved by erosion after these rock layers were tilted steeply by uplift of the present Rocky Mountains. The concentration of monoliths at the site is probably mostly accidental, though the rocks may be cemented just a little better here than elsewhere and so be more resistant to erosion. There is a similar group of rocks three miles south in Roxborough Park, and other more or less isolated monoliths rise elsewhere along the east side of the Front Range.

Red and white concentric bands on the rock surfaces cut across both bedding planes and cross-bedding striations, especially on the north wall of the Amphitheatre. They are caused by solution and chemical changes in the iron mineral **hematite** that gives the rock its color and the park its name. The banding formed while the Fountain Formation was buried under thousands of feet of sediment, as groundwater seeping through pores in the rock caused chemical changes. Notice also the rough, corrugated surfaces of the monoliths. Sandstone layers are harder and more resistant to erosion by wind and water than siltstone layers, so they tend to project and the siltstone to recede, providing good examples of **differential erosion**.

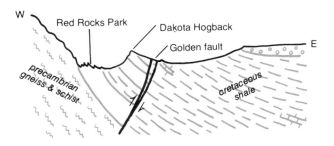

East-west section across Red Rocks Park

At the west edge of the park Precambrian metamorphic rocks rise steeply west of the Pennsylvanian red sandstones. With one hand you can span the contact between the two kinds of rock. In so doing your hand spans a time interval of nearly 1.5 billion years, from the Precambrian Era to the Pennsylvanian Period! Such a contact, with an interval of time unrecorded or recorded in rocks that are later destroyed, is an **unconformity**. This one represents one of the most widespread and longest periods of erosion the Earth has known. Two centuries ago James Hutton, "Father of Geology," gave the name "Great Unconformity" to a corresponding feature in Scotland.

The long ridge of the Dakota Hogback rises east of Red Rocks Park. Bordered on the east by soft Lower Cretaceous shale, and on the west by equally soft Jurassic shale, this hogback stands out (as do the Red Rocks themselves) as a much larger example of differential erosion. The Interstate 70 highway cut two miles north of here lays bare the layered rocks, particularly the soft and easily eroded ones, much better than natural outcrops, but they are fairly well exposed also at Morrison, along Alameda Parkway, and at the roadcut for U.S. 285.

The Dakota Hogback forms a curving arc along the east side of the Front Range, partly enclosing Red Rocks Park, close to the mountains in the left center. There, the Fountain Formation tips up right at the mountain front; hills left of it are Precambrian gneiss and schist. A small hogback of Cretaceous limestone catches the sunlight in the right foreground. T.S. LOVERING PHOTO, COURTESY OF USGS

Jurassic shale and sandstone between Red Rocks Park and the Dakota Hogback, forming also the west slope of the hogback, belong to the Morrison Formation, named for the little town at the southeast corner of Red Rocks Park. Jurassic dinosaur skeletons unearthed here between 1876 and 1878 gave international fame to both the rock formation and the nearby settlement. Several entire skeletons were dug out and moved to eastern museums for study and exhibit, including remains of the largest "thunder lizard" then known, *Brontosaurus*. The Morrison Formation was deposited at a time when Colorado was a flat and low-lying coastal plain with low bars and beaches, marshes and tidal flats. Eventually the sea advanced across this part of the continent, coming from the east and bringing with it near-shore sands of the Dakota Formation and later fine black muds, now turned to Pierre Shale, found east of the Dakota Hogback.

Remember that all these rock layers were deposited in horizontal sheets, and are here bent up by the uplift of the mountains. Rocks are more flexible than we normally give them credit for, especially when there is a lot of time on hand so that they can bend slowly. But they do

77

develop many **joints** (cracks) when they are under stress.

Extensive faulting here dates from the Laramide Orogeny and possibly also from the later regional uplift. The largest of the faults is the Golden fault, which runs east of the Dakota Hogback and is well concealed by overlying shales and soils. There has been greater movement along it than along most other faults on the east flank of the Front Range: matching the rocks on opposite sides of it shows that displacement is nearly 11,000 feet. It swings over toward the Precambrian mountain core near Golden, cutting out the Dakota Hogback and some other Cretaceous rocks there, as well as Jurassic, Triassic, and Permian rocks, and part of the Pennsylvanian red rocks that correspond with those at Red Rocks Park.

mount evans

(118-mile round trip from State Capitol)

The road up Mt. Evans climbs from 7500 feet at Idaho Springs to a point well above 14,000 feet. A short trail then leads to the summit, 14,264 feet above sea level. Take it easy on the trail if you are not used to climbing at high altitude.

Mt. Evans is made of granite that pushed and melted its way into overlying metamorphic rocks as a huge intrusion of molten magma, a **batholith**, 1700 million years ago, during one of the mountain-building episodes in Precambrian time. Having cooled extremely slowly, the granite is coarse-grained, and you can easily pick out the minerals of which it is made: glassy quartz, white and pink feldspar,

Quartz veins often form intricate patterns in Precambrian gneiss and schist. These are on Grays Peak, not far from Mt. Evans.

T.S. LOVERING PHOTO, COURTESY OF USGS

78

and black mica or biotite. Along with surrounding older metamorphic rocks that originated in still earlier mountains, the granite was faulted and pushed upward 300 million years ago as part of the Ancestral Rockies, and again 65 million years ago during the Laramide Orogeny as part of the Front Range.

In the stresses and strains of these many episodes of mountain-building, the rock cracked and broke and was impregnated with mineral liquids, so that now it is cut by myriads of joints and criss-crossed with narrow white veins of milky quartz. Each parallel set of joints probably records a different stress to which the granite was subjected during its 1700-million-year history. At the surface, weathering has widened most joints and in many cases rounded the blocks into separate boulders in a process called **spheroidal weathering**. Characteristic of granite, this type of weathering occurs because mica grains change gradually into clay when exposed to water. The grains enlarge during this change, and their increase in size causes internal pressures that pop off individual loosened quartz grains and even whole surface layers. Weathering is most rapid at corners and sharp edges where the rock is exposed on more than one side, so gradually the boulders become more and more spherical. Around the base of the round boulders formed by this process you can often see coarse quartz and feldspar sand — the first step in the making of soil.

Above timberline, where geologic features are no longer hidden by trees, many geologic details such as joints, faults, and **exfoliation** — peeling off of outermost curved layers of rock — can be clearly seen in the granite. **Wind blast pitting** forms small smooth hollows on the western windward side of some exposed rocks, especially in the Alpine Garden area and along the trail to the summit. Watch for veins of milky white quartz and for **pegmatite dikes**, bands of granite-like rock with very large crystal grains and often very light color.

Summit Lake lies in a small glacier-scoured basin that is part of a horseshoe-shaped cirque carved by glacier action in the hard gray granite of the mountain. Walk the short trail to the low divide beyond the lake to look down 2000 feet into the magnificent cirque and valley of Chicago Creek. Valleys like this one, U-shaped in cross section, are hallmarks of mountain glaciation, for streams carve distinctly V-shaped valleys. A string of **paternoster lakes**, so-named because they are strung down the valley like beads on a rosary, decorates the valley floor, which is sculptured in step-like levels also typical of glaciated mountain valleys. Rock debris piled by a glacier into a **terminal moraine** dams the lowest lake. Straight across the valley, falling rock debris forms long talus cones below exposed cliffs. Where do the broken rocks come from? When rain or melted snow trickles into joints and cracks and crevices, and then freezes, the expansion due to freezing gradually wedges the rocks apart.

The summit of Mt. Evans offers a beautiful panorama of the Front Range from Longs Peak, 50 miles due north, to Pikes Peak 58 miles away to the south-southeast. These peaks, as well as Mt. Evans itself and nearby Mt. Bierstadt and Torreys and Grays Peaks, are "Four-teeners," over 14,000 feet high. They stand 9000 feet above the Colorado Piedmont and about 5000 feet above the sloping Tertiary pediment, which can be clearly seen from here, particularly in the Rampart Range between Mt. Evans and Pikes Peak.

Far to the south lie the northern peaks of the Sangre de Cristo Range, a young fault block range that stretches south to Santa Fe, New Mexico. To the southwest is South Park, one of Colorado's four large intermontane basins. Nearer than South Park U.S. 285 winds across Kenosha Pass. And westward the three highest summits are Torreys and Grays Peaks and Mt. Bierstadt.

The complexity of the geology in this area is hard to believe. The Precambrian rocks have been mapped several times, but the geology is so complex and the outcrops are so scattered that each geologist who works on them makes a map slightly different from those of others.

Boulder's Flatirons are composed of sandstone and conglomerate deposited as alluvial fans and aprons along the edges of the Ancestral Rockies. They are now bent up sharply at the edge of the Front Range.

boulder-estes park-central city loop
on u.s. 36, co. 7, 72 and 119
and u.s. 6
(150 miles)

U.S. 36 angles northwest from Denver across Tertiary and Cretaceous sediments that fill the Denver Basin. Stop briefly at the scenic overlook a few miles before reaching Boulder, on top of a fault block of Upper Cretaceous sandstone. Boulder Valley, like much of the Colorado Piedmont, is floored with Pierre Shale, also Cretaceous. Under the shale are older flat-lying Mesozoic and Paleozoic sedimentary layers — a total of about 10,000 feet of them. Where not removed by faulting they rise to the surface along the mountain front, but only the erosion-resisting ones are clearly exposed.

The most striking sedimentary rocks are the Flatirons, the great rocks that lean against the mountains west of Boulder. Pennsylvanian red sandstone and conglomerate deposited along the edge of the Ancestral Rocky Mountains — the same rocks that appear in Red Rocks Park west of Denver — lean directly on Precambrian granite; older Paleozoic rocks were washed off the Ancestral Rockies before the Pennsylvanian ones were formed. They now have been dragged upward by the rebirth of the Rockies during the Laramide Orogeny. Erosion of small mountain canyons and removal of softer layers created the unusual shape for which the Flatirons are named.

Mesas south and north of Boulder are parts of mountain pediments eroded in upturned Cretaceous sedimentary rocks and later covered with 10 to 20 feet of younger gravel. Though scarcely consolidated, gravel makes a caprock because its permeability inhibits runoff and therefore prevents erosion. Beyond the pediments the Dakota Hogback borders the mountains; near Boulder Creek it has been cut by faults.

The Tertiary pediment is quite evident from the overlook, halfway up the Front Range.

81

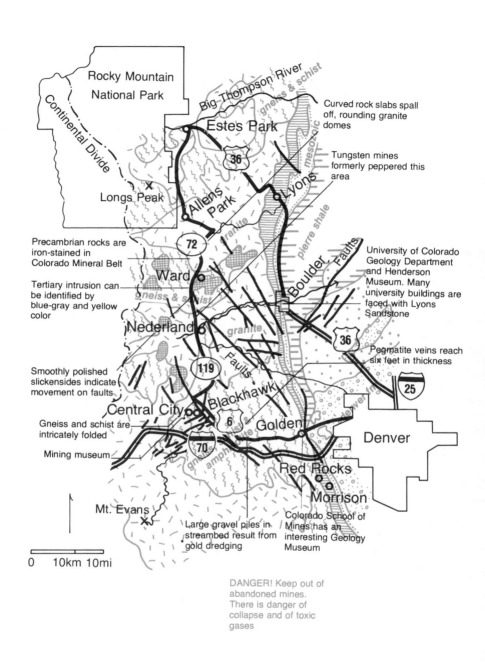

Rocky Mountain National Park

Continental Divide

Big Thompson River

gneiss & schist

Estes Park

36

Longs Peak

Allens Park

72

Lyons

granite

mesozoic

pierre shale

Curved rock slabs spall off, rounding granite domes

Tungsten mines formerly peppered this area

University of Colorado Geology Department and Henderson Museum. Many university buildings are faced with Lyons Sandstone

Precambrian rocks are iron-stained in Colorado Mineral Belt

Tertiary intrusion can be identified by blue-gray and yellow color

Ward

gneiss & schist

Nederland

granite

Boulder

Faults

36

Smoothly polished slickensides indicate movement on faults

Faults

119

Pegmatite veins reach six feet in thickness

25

Blackhawk

Central City

6

Golden

Denver

Gneiss and schist are intricately folded

Mining museum

70

gneiss

amphibolite

denver fm.

Red Rocks

Morrison

Mt. Evans

0   10km 10mi

Large gravel piles in streambed result from gold dredging

Colorado School of Mines has an interesting Geology Museum

DANGER! Keep out of abandoned mines. There is danger of collapse and of toxic gases

Mt. Evans and Boulder-Estes Park-Central City loop

North of Boulder U.S. 36 parallels the Dakota Hogback, running on the edges of upturned Mesozoic sedimentary rocks that sometimes make smaller, less continuous limestone and sandstone hogbacks. Steeply dipping dark gray Pierre Shale shows on mesa slopes east of the road. Marine fossils occur in the limestones as well as in the Pierre Shale: clams, oysters, occasional bones of extinct swimming reptiles, and tiny one-celled hard-shelled Foraminifera. South of Lyons the limestones are quarried for cement.

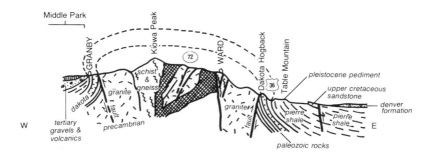

Section across U.S. 36 and Colorado 72, showing the structure of the Front Range. The great block of Precambrian rock expanded as it was pushed upward, so that the faults on either side are reverse faults with the upward-moving block overhanging sedimentary rocks on either side. Dotted lines show hypothetical anticline in Paleozoic and Mesozoic rocks.

Close to Lyons the highway turns abruptly west, going through the Dakota Hogback and into progressively older rocks: valley-forming Jurassic shale and more resistant Triassic and Permian sandstone. Fine, even-grained, salmon-colored Permian rocks are quarried north of Lyons for flagstone and building stone, for they have an attractive color and a convenient tendency to break smoothly along cross-bedding surfaces that run diagonally to the real bedding. The cross-bedding is of a type that indicates this rock formed as an ancient dune deposit, and the footprints of primitive four-footed creatures have been found in the quarries here.

Northwest of Lyons, U.S. 36 plunges westward into the Precambrian granite of the mountain core. A part of a batholith, the granite formed as a mass of slowly cooling molten rock or magma, and was part of a Precambrian mountain system. It is made up of small chunky crystals of glassy quartz, pink and light gray feldspar, and black mica. Sometimes it changes gradually into gneiss, banded strongly in shades of black and gray and pink. Schist also occurs here, with flat-faced mica crystals causing it to break in platy fashion. All three of these rock types are cut by pegmatite dikes (bands of coarser, larger crystals) and by white quartz and feldspar veins. Dated by radioactivity analysis, the gneiss and schist turn out to be 1,750 million years old, and the granite 1,450 million years old.

Near an overlook between Miles 3 and 4, an unusual type of gneiss is exposed — **augen gneiss**, from a German word for eyes. The eyes are centered with crystals of iron-rich black mica; the areas around the augen, having given up their iron to the centers, appear lighter, like the whites of eyes.

The valley of Big Thompson Creek, below the overlook, was probably carved during glacial times when rainfall, snowfall, and stream runoff were far greater than now. A deep stream-carved canyon underneath the present valley floor is filled in with debris washed from glaciers higher up the valley, above 8,000 feet elevation. Estes Park is a taking-off point for trips into Rocky Mountain Park, a fine place for seeing the effects of mountain glaciation. The canyon of Big Thompson River below (east of) Estes Park was gutted by a flash flood in 1976, with 139 lives lost, and also makes an interesting trip (see U.S. 34 LOVELAND — GRANBY).

Turn south on Colorado 7. This highway climbs through more Precambrian granite, gneiss, and schist, and in about eight miles levels off onto rolling, fairly open country about 9000-10,000 feet in elevation — the Tertiary Pediment. The route remains on this surface almost to Blackhawk, dipping down only to cross major mountain stream valleys.

Views of Long's Peak, west of Colorado 7, show the "Diamond" of the peak's east face and tree-covered lateral and terminal moraines. Trails from Longs Peak Campground lead to the cirque and to the summit, 14,255 feet above sea level. JACK RATHBONE PHOTO

84

Longs Peak's towering 2000-foot sheer granite face was carved by glaciers in Pleistocene time. Banded metamorphic rocks appear on the foreground ridge. Chasm Lake occupies an ice-scooped cirque.

JACK RATHBONEPHOTO

Magnificent views of Longs Peak to the west show the 2000-foot sheer granite cliff of the peak's east face, the headwall of a glacial cirque. Some geologists think the flat top of Longs Peak is a very old peneplain surface corresponding to the surface of Precambrian rocks at the end of the Lipalian Interval, 600 million years ago. The top of the Precambrian rocks on Longs Peak is 22,000 feet higher than that in the deepest part of the Denver Basin, so we know that the mountains have risen at least four miles!

Ward and Nederland mark the north end of the Colorado Mineral

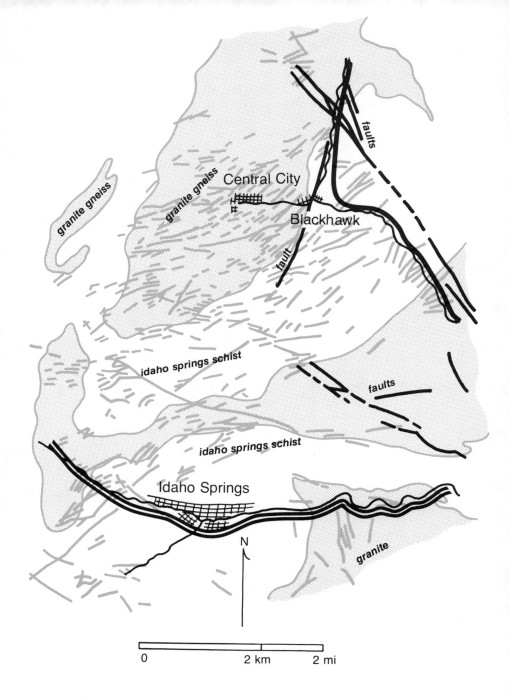

Hard though it may be to believe, this is a much simplified map of the Central City-Idaho Springs mining district, showing only the three main rock types, Idaho Springs Schist, granite, and granite gneiss. The more than 500 known mineral veins (most of them now mined out) are shown in color

This 1917 photo shows the tungsten mill that used to operate just below the present Barker Dam in Boulder Canyon. The nearby town of Ferberite, no longer in existence, was named after a tungsten ore mineral. F.L. HESS PHOTO, COURTESY OF USGS

Belt, a 50-mile-wide zone that extends from here to southwestern Colorado. This belt contains most of the mines of the state. Many of the ore deposits are near boundaries of igneous intrusions, where during the Laramide Orogeny hot solutions emanating from the intrusions penetrated cracks and crevices in the surrounding rocks. Gold, silver, lead, copper, zinc, molybdenum, uranium, and tungsten have been mined in this district. Along the highway just south of Ward there is a fine-grained gray Tertiary intrusion that may be related to the mineralization there.

The Central City-Idaho Springs mining district has produced almost $200,000,000 worth of gold, silver, lead, zinc and copper. In 1859 prospectors who worked placer deposits in the streambed gravels near Golden made their way upstream searching for the source of the placer gold. One named John Gregory struck it rich in Gregory Gulch between Blackhawk and Central City. Mines dot the slopes above these towns in the "richest square mile on earth." Gold occurred in vertical or nearly vertical quartz veins that surrounded the town and extended southwest through the mountains toward Idaho Springs. The veins filled myriads of cracks and fissures in shattered Precambrian rock; locate them by looking for collapsed tunnels, for the ore-bearing rock has been mined out completely. Mine tours and jeep tours are available here.

South of Blackhawk, Colorado 119 and Clear Creek (from which a bit of "color" can still be panned) follow the weakened rock along the Blackhawk fault zone. Most of the rock in Clear Creek Canyon is either gneiss or schist. Amphibolite, a dark gray metamorphic rock in which the primary mineral is hornblende, occurs as well; it lacks schist's platy, flaky way of breaking, so the two are easy to tell apart even though they are both dark gray.

At the mouth of Clear Creek Canyon near Golden the route leaves the Precambrian core of the Front Range. Here, along the large fault zone on the east side of the Front Range, most of the Paleozoic and Mesozoic sedimentary rocks that in most places edge the range have been cut out. So there are no Flatirons or Dakota Hogback. Instead, two flat-topped mesas frame Golden. They are made of tilted Cretaceous and Tertiary sedimentary rocks protected from erosion by flat lava caps. You can see the vertical **columnar jointing** in the lava flows. Plant fossils occur in soil zones between the layered volcanic flows, and dinosaur bones and tracks have been found in the Cretaceous rocks.

colorado springs area

Originally the town of Colorado Springs nestled near Monument and Fountain Creeks, between the southern end of the Front Range and a cluster of sheltering white sandstone buttes and pinnacles. The city has grown immensely, overflowing eastward across the sandstone hills and creeping up westward onto pediment terraces that project like paws of giant sphinxes from the edge of the mountains.

The old part of town lies on dark gray Pierre Shale, a fossil-bearing Cretaceous rock laid down as the muddy bottom of a sea that stretched from the Arctic to the Gulf of Mexico 75 million years ago. Here the shale begins to reflect the upward drag of the mountain mass, as it tilts up westward toward the mountains.

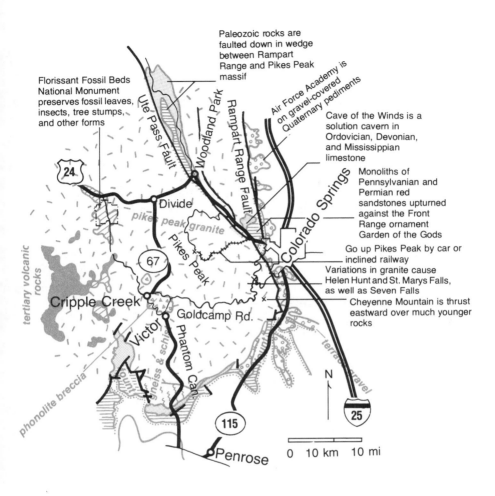

Florissant Fossil Beds National Monument preserves fossil leaves, insects, tree stumps, and other forms

Paleozoic rocks are faulted down in wedge between Rampart Range and Pikes Peak massif

Air Force Academy is on gravel-covered Quaternary pediments

Cave of the Winds is a solution cavern in Ordovician, Devonian, and Mississippian limestone

Monoliths of Pennsylvanian and Permian red sandstones upturned against the Front Range ornament Garden of the Gods

Go up Pikes Peak by car or inclined railway

Variations in granite cause Helen Hunt and St. Marys Falls, as well as Seven Falls

Cheyenne Mountain is thrust eastward over much younger rocks

Ute Pass Fault

Woodland Park

Rampart Range Fault

Colorado Springs

24

Divide

pikes peak granite

Pikes Peak

67

tertiary volcanic rocks

Cripple Creek

Goldcamp Rd.

Victor

gneiss & schist

Phantom Can.

phonolite breccia

N

terrace gravel

115

25

Penrose

0   10 km   10 mi

Colorado Springs area. Many known faults are deleted.

A network of particularly large faults separates the mountains from the city, chief among them the Ute Pass and Rampart Range Faults. The Rampart Range Fault trends north-south, edging the mountain uplift north of Fountain Creek and separating the Rampart Range from the sphinx-paw pediments of the northern part of the city and of the Air Force Academy grounds. The Ute Pass Fault fronts Cheyenne Mountain southwest of the city, separating Pikes Peak and Cheyenne Mountain on the west from more pediments and the valley of Fountain Creek on the east. This fault bends at Manitou and runs northwest into the mountains, separating Pikes Peak with its 14,109-foot elevation from the Rampart Range, only about 9000 feet in elevation. U.S. Highway 24 northwest to Woodland Park

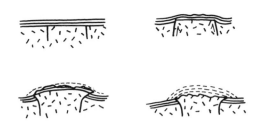

As the granite mountain core rose, it expanded and spread over surrounding sedimentary rocks. Erosion bit into the top of the block as soon as uplift started, of course, and eventually completely removed all traces of sedimentary rocks that arched over the uplift block.

follows the broken, weakened rock of the fault zone up the fault-controlled course of Fountain Creek.

At Manitou Springs, mineral waters bubble to the surface through a number of channels in the shattered rock along Ute Pass Fault. These naturally carbonated waters have absorbed carbon dioxide from carbonate rock — Paleozoic limestone — at some depth, where pressure is similar to that in a capped pop bottle. As the water rises rapidly along the Ute Pass fault zone, pressure on it decreases — just as it does when you uncap a bottle of pop — and the carbon dioxide comes out of solution to form the bubbles that give Manitou water, and soft drinks, their refreshing "fizz."

As the great granite mass of the Pikes Peak massif rose during the Laramide Orogeny, the granite was to some extent released from the inward pressures caused by surrounding rock. Though it is hard to believe that what seems eternally solid, like crystalline granite that cooled and solidified a billion years earlier, can actually expand like toothpaste coming out of a narrow tube, the Pikes Peak Granite appears to have done just that, spreading outward horizontally over surrounding sedimentary rocks. Along the base of Cheyenne Mountain, the granite has ridden out about a mile over sedimentary rocks. Of course in the whole mountain-building process the granite was extensively broken, but some of it seems to have actually expanded uniformly as a sum total of molecular expansion in its uncounted billions of individual mineral crystals. The same phenomenon occurs in other parts of Colorado, where vertically rising, horizontally spreading blocks are edged by thrust faults. Where this happens, it plays havoc of course with geology's usual youngest-on-top sequence. In these places the ancient granite or metamorphic rock, sometimes more than a billion and half years old, ends up on top of sedimentary rocks whose age is measured in mere hundred-million-year units.

garden of the gods

In the narrow point of land between the two major faults that edge the mountains near Colorado Springs, a wedge of Paleozoic and Mesozoic rocks has been preserved. Bent up by the mountain uplift and cut by many lesser faults, these rocks remain as spectacular towers and pinnacles and mushroom rocks of Garden of the Gods, and as shale and limestone ridges farther east and west.

The road entering Garden of the Gods from the east crosses a broad swale underlain by soft Cretaceous shale and a small wall-like hogback of white rock that is a resistant sandstone layer in the Jurassic Morrison Formation. Both these rock units are tilted into an almost vertical position. West of another swale, one eroded in soft red gypsum-bearing Triassic shale, the road slips through the narrow gateway to Garden of the Gods, between two gigantic gateposts of salmon-colored beach and shore sandstone of probable Permian age. Here the sandstone seems a lot more resistant than elsewhere, probably because groundwater carrying mineral "hardeners" moved more freely along a nearby fault. The sand grains in this rock are cemented together by silica and to some extent by the mineral hematite, which gives the rock its pinkish color. Tilted until they are vertical — or in fact slightly overturned so the monoliths lean eastward a little — the sandstone layers are eroded into a succession of tall, thin, blade-like towers.

From the Visitor Center and viewpoint you can see that these rocks are not all restricted to the Garden of the Gods area. They continue south in regimental rows across Fountain Creek until they are cut off abruptly by the Ute Pass Fault, with the steep slopes of Cheyenne Mountain rising behind.

Farther west in Garden of the Gods the rocks are darker red, and the sandstone alternates with deep red mudstone or shale and sometimes with bands of pebbly conglomerate. Here the rocks are probably Pennsylvanian in age. They do not dip as steeply, and are shaped into more modest forms, mushroom-like pedestals and "balanced" rocks that result from differential erosion of harder and softer layers. A fault zone concealed by the soil of the valley floor separates them from the tall gateway slabs farther east.

All of these pink and red rocks were deposited originally around Frontrangia, the eastern range of the Ancestral Rockies, which if you remember rose during the Pennslvanian Period parallel to but a little farther west than the present Front Range. Mountain streams rushing down the slopes of these ancient mountains spread rock debris around them in great alluvial aprons, which have since turned into these colorful monuments. Look closely for sedimentary details in these rocks — ripple marks, thin bands of rounded stream-worn pebbles, sometimes the impressions of mudcracks which show that the mud, now rock, was deposited where it could occasionally dry out. In places you can see the sweeping lines of cross bedding cutting diagonally across sandstone layers.

We do some guessing as to the Pennsylvanian and Permian age of Garden of the Gods rocks. Actually educated guessing for we know they were deposited above Pennsylvanian shales and below Triassic ones. But their lack of fossils makes it impossible to date them exactly. The lower part, at the west side of Garden of the Gods, we assume to be Pennsylvanian because some of the lowest layers alternate with thin gray shales that do contain Pennsylvanian fossils. The upper portions, including the tall pink sandstone pinnacles, look like the Permian Lyons Sandstone north of Boulder, but except for a few animal footprints in the sand surfaces, that rock contains no fossils either.

cave of the winds

Cave of the Winds near Manitou is the only limestone cavern in the state developed as a tourist attraction. It formed where groundwater, made slightly acid by passing through forest soils and the granite of the mountains, moved more easily than usual along abundant joints and fissures at the edge of the mountain uplift. In layers of Paleozoic limestone the groundwater sought always the easiest downward route, and as it gradually dissolved and washed away some of the limestone, it formed the many rooms and passageways of the present cavern. The cave was probably shaped in Ice Age (Pleistocene) time, when both surface runoff and groundwater were much more abundant than at present.

As groundwater supplies decreased at the end of the Ice Age, solution decreased too, and eventually water in the cave was reduced to the mere trickles that you see today. Where dripping lime-laden water evaporated, **stalagmites** and **stalactites** developed, for each evaporating droplet left behind the tiny bit of the mineral calcite it held in solution. The cave's stalagmites and stalactites and their curtain-like and ribbon-like variations were probably deposited during just the last few thousand years. The curtains in particular, and some rows of stalactites, line up along joints or cracks in the ceiling rock.

*Stalactites (with a **c**) hang from the ceiling. Stalagmites (with a **g**) rise from the ground.*

These features develop slowly, and any damage to them now seems to last forever, for as you can see there is little moisture in the cave at present and without it nature can do no repair work.

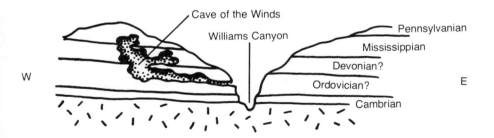

The exit road passes through Williams Canyon, where the limestone in which the cave is dissolved is exposed by the roadside. At the edge of the canyon floor you can see Cambrian sedimentary rocks lying almost horizontally on coarse pink Precambrian granite, separated by a smooth surface beveled during the long Lipalian Interval at the end of Precambrian time. Above them and forming the canyon narrows is massive gray Ordovician limestone in which the deepest level of Cave of the Winds is carved. This rock contains many fossils, in particular trilobites, brachiopods, and straight-shelled or horn-like cephalopods, ancient ancestors of today's Chambered Nautilus.

The Ordovician limestone has above it layers of limestone of uncertain age — probably Devonian — that contain few fossils — only tiny teeth-like Conodonts. This is the rock that forms middle-level rooms and passageways in Cave of the Winds. Limestone of the uppermost parts of Cave of the Winds lies above that. Here also are occasional fossils enabling us to identify its age as Mississippian for sure. Pennsylvanian marine shales, some of them quite fossiliferous too, occur higher on the slopes east of Williams Canyon.

In Williams Canyon, northwest of Manitou, Cambrian Sawatch Sandstone and Ordovician Manitou Limestone lie on Precambrian granite, barely visible in the streambed. The lowest parts of Cave of the Winds are within the Manitou Limestone, behind the cliffs in the left part of the photograph.

N.H. DARTON PHOTO, COURTESY OF USGS

Pikes Peak (elevation 14,109 feet) is the southern end of the core of the faulted anticline that forms the Front Range. Red monoliths in the Garden of the Gods rise in the foreground.   JACK RATHBONE PHOTO.

pikes peak

(56-mile round trip)

Pikes Peak is made entirely of Pikes Peak Granite, a beautiful pink granite that contains stubby interlocking crystals of glass-like quartz and flat-faced white and pink feldspar, with a liberal sprinkling of hornblende and black flaky mica. About a billion years old, this ancient rock developed as a batholith, an immense mass of hot molten rock or **magma** that pushed and melted its way upward through the earth's crust. It never made it to the surface, but cooled deep down and very slowly over probably millions of years.

Doubtless once the core of a Precambrian mountain range, beveled during the Lipalian Interval and later covered with thin layers of Paleozoic sedimentary rock, the granite was pushed up again — not molten this time — to form part of the Ancestral Rocky Mountains. Then as those highlands were attacked by erosion the granite was

95

eventually beveled once more, and later covered with new thicker layers of sediment of later Paleozoic and Mesozoic age. Uplifted for the third time during the Laramide Orogeny, it was stripped once more of overlying sediments and today is slowly yielding anew to the processes of erosion.

One of the first steps in the erosional process can be seen easily from the road up the mountain: the weathering of granite and its gradual conversion to soil. Mica crystals in the granite slowly decompose, especially when exposed to moisture, changing into clay minerals. But the clay minerals take up more space than the previous micas, swelling and loosening the quartz crystals that form much of the body of the rock. The swelling process is augmented by frost action and by expansion of water that seeps into cracks between crystals and, on cold nights, turns to ice. The process gradually weakens the granite, forming a deep layer of soft "rotten" granite, the first step in the making of soil.

Weathering of course takes place faster where there are joints and crevices in the rock, and the Pikes Peak Granite, thanks to its checkered billion-year history, has plenty of these. Often weathering of jointed rock forms separate rounded boulders, for weathering attacks sharp corners from two sides at once.

As the road climbs higher on the slopes of Pikes Peak, you will begin to see features left by the glaciers that existed during the frigid years of the Pleistocene Ice Ages. The spacious cirques of Bottomless Pit and the so-called Crater (a real misnomer because the mountain is definitely not a volcano!) were carved by large valley glaciers that ground down the mountainside, scouring steep-walled U-shaped canyons that extend as low as 8000 feet, the level at which the downstream ends of the glaciers melted.

From the high slopes you can look out, too, over the Tertiary pediment's nearly horizontal surface, sprawling westward some distance and all the way north to the slopes of Mt. Evans west of Denver.

At overlooks near the top of the mountain the granite is so intricately jointed that it looks as though someone had chopped it with a meat chopper. Several crisscrossing sets of parallel joints can be recognized, each representing a different direction of stress to which the rock was at some time exposed.

There is less weathering here because moisture is often in the form of snow, which doesn't sink readily into the rocks to break down the mica crystals. But moisture from snowmelt seeps into the joints,

In Pikes Peak granite, as well as in other massive rocks, weathering follows joint planes, separating boulders and rounding their protruding angular edges.

T.S. LOVERING PHOTO, COURTESY OF USGS

where it freezes and wedges the rocks apart. Such fracturing of jointed rock near the summit produces a barren jagged surface with angular boulders tipped at crazy angles — a boulder field resulting from years and years of frost action. Small particles of rock have been whipped away by violent mountain winds. Here, too, weathering is slow, though it is somewhat helped along by acids secreted by the lichens that grow on some rock surfaces.

In 1806 President Jefferson dispatched an expedition to explore the Pikes Peak region, part of the newly purchased Louisiana Territory. The party was led by Zebulon Pike, who attempted to climb to the top of the mountain that he called the "Grand Peak" and that he estimated was 18,581 feet high. Mapped first as James Peak, it was commonly called Pike's Peak by trappers and military men, and Pike's Peak it became. The U.S. Geological Survey now customarily leaves out the apostrophe in names like Pikes Peak, Longs Peak, and St. Marys Falls.

The name was for a time used for the entire Front Range region, leading to the "Pikes Peak or bust" slogans of gold-rush miners heading for the goldfields near Denver. Gold was not discovered near Pikes Peak until 1891, 32 long years after it was found in the canyons west of Denver. The mountain is said to have been the inspiration for the anthem "America the Beautiful," with its "Purple mountain majesties / Above the fruited plain."

The town of Independence, in the Cripple Creek area, was dotted with prospect puts and mine dumps. By 1903, when this picture was taken, most of the native forest had been cut for fuel and mine timbers.

F.L. RANSOME PHOTO, COURTESY OF USGS

cripple creek
and gold camp road
(80-mile loop)

For the first part of this route, follow the itinerary for U.S. 24 COLORADO SPRINGS — BUENA VISTA. At Divide turn left on Colorado 67 to Cripple Creek, or continue for four more miles on U.S. 24 and then turn left, taking in FLORISSANT FOSSIL BEDS NATIONAL MONUMENT on the back road to Cripple Creek.

The Cripple Creek mining district, on the west side of Pikes Peak, was discovered in 1891, fairly late in Colorado's mining history. The district produced close to $450,000,000 in gold and silver, with mining activity peaking around 1900. Increasing prices for gold and silver are causing a renewal of activity in this area, and a number of mines have reopened.

Cripple Creek mining district is an ancient volcanic caldera cut by myriad faults and veins.

The ore deposits are all in a small area four miles long and two miles wide, between Cripple Creek and its early rival, Victor. They are in Tertiary volcanic rocks surrounded by pink Precambrian intrusive and metamorphic rocks.

The mineralized area is a collapsed volcano, a **caldera**. A mass of volcanic and non-volcanic rock fills the caldera, a mixed-up mash of angular rock fragments to which geologists give the Italian word **breccia**, meaning broken. The rock seems to have been shattered and cemented and shattered again several times, suggesting that the caldera collapsed repeatedly at the same site and was filled with sediment and new volcanic rocks between collapses. It is cut by many dikes and several irregular tubes of volcanic material, sometimes black basalt but more often a lighter rock called **phonolite**. This word is a "Greekified" version of the more picturesque miners' term **clinkstone**, which refers to the clinking sound the rocks make when struck with a hammer.

In periods between collapses, mineral-rich fluids seeped into the cracks and crevices in the caldera and hardened into ore-bearing veins. In addition to gold and silver ores the veins contain crystals of fluorspar, pyrite, galena, calcite, and other minerals. Many of these minerals can be found on mine dumps, but **keep away from old mine shafts and tunnels, for they may harbor toxic gases and often they are on the verge of collapse.** Most of the richest veins are around the fringes of the breccia mass; some extend into it or into the surrounding granite.

Three narrow-gauge railroads vied for Cripple Creek business in the mining heyday. All three were dismantled when the glory faded. One was along the route of Colorado 67, which you may have followed between Divide and Cripple Creek. Another came north from Florence. The third, now called Gold Camp Road, skirts the south flank of Pikes Peak, penetrating ridges and mountain spurs with cuts or tunnels just wide enough for the old narrow-gauge railway engines and cars. At an altitude just below 10,000 feet, much of the road is on the surface of the Tertiary pediment.

Along this route, examine weathering characteristics of the Pikes Peak Granite. In places it is cut by many joints, usually arranged in parallel sets — one set vertical, another horizontal, still another vertical but running in a different direction. Because sharp, angular edges are attacked by weathering processes on two sides at once, such fractured granite tends to end up as rounded boulders. (See PIKES PEAK for a further explanation of this process.) At Cathedral Rocks such rounding has created huge rocky turrets and pinnacles.

Eventually the granite decomposes completely into coarse sand containing feldspar and quartz grains about the same size as the original crystals in the granite. With the addition of fallen leaves, grass roots, and other organic matter, the sand becomes loose forest soil. Many steps in the soil-making process can be seen along this route.

At Devils' Slide and St. Peters Dome, where the granite isn't so closely jointed, the same type of weathering has flaked off huge curving sheets of rock. Even rock as stiff and unyielding as granite expands, at least a tiny bit, when it is freed from the weight of overlying rock, so a set of joints caused by expansion parallels the large exposed surfaces. Weathering of mica crystals along these joints loosens the surface layer, and water that has seeped into the joints freezes by night and melts by day, further widening the once tiny cracks. Finally the surface layer loosens and spalls off as an oversized rock flake.

Helen Hunt Falls and St. Marys Falls, on the east side of Pikes Peak, occur where a band of more resistant, less jointed granite runs across much more heavily jointed granite that is more easily eroded by the processes described above. Here too one can see the typical rounded weathering of granite, and notice that parts of the forest floor are composed almost entirely of coarse quartz and feldspar grains loosened by weathering of mica crystals.

Great arcuate slabs peel off large masses of granite; the joints which separate them result from expansion of granite as it is relieved of confinement pressures.
T.S. LOVERING PHOTO, COURTESY OF USGS

102

iv
purple mountain's majesty — folded and faulted ranges

What a frightening barrier Colorado's majestic mountains must have seemed to earlier westward travelers with their lumbering ox-drawn wagons! Following the great western routes — the Santa Fe Trail, the Oregon Trail, and others — they turned aside, preferring the low passes of New Mexico and Wyoming to the threatening Colorado peaks. Even now only a handful of highways challenge the high passes, where the traveler gazes in awe at midsummer snowfields and wind-swept tundra. In winter, avalanches threaten.

Study of rocks in and along the flanks of the dozen or so ranges of the Colorado Rockies reveals that most of them are giant corrugations in the Earth's crust, formed where Precambrian rocks that seem to make up the ancient "basement" of all the continents were broken into long north-south slices and raised thousands of feet. Younger, more flexible layers of sedimentary rock, once flat-lying above the basement, were lifted too. Instead of breaking they often stretched and draped like rippled blankets across the broken edges of the mountain cores.

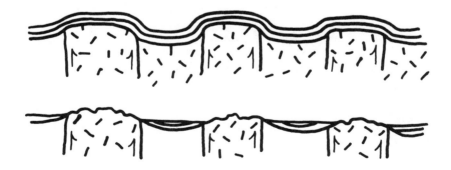

Mountains of this kind are termed **faulted anticlines,** faulted because of the broken basement rocks underneath, and anticlines because of the upward arched layered rocks along their margins.

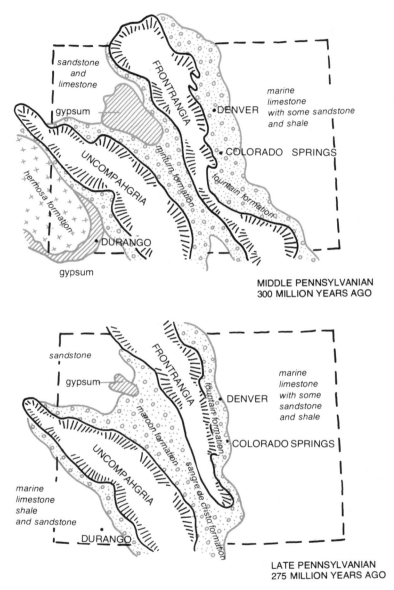

Frontrangia and Uncompahgria, two mountainous islands in the sea, rose in Pennsylvanian time. Material eroded from them encircled them with alluvial aprons, now the Fountain and Maroon Formations. In evaporative basins, salt and gypsums were deposited.

Colorado's present mountains rose early in the Cenozoic Era (Age of Mammals), starting about 65 million years ago. The uplift itself actually went on sporadically for 20 million years or more, for mountains are not made overnight. An earthquake here, a shifting fault there, 20 feet, 5 feet, perhaps only inches at a time, multiplied by 20 million years. The same process goes on today in Chile and Turkey and other regions where the Earth's crust is restless still.

But let's go back farther in time; the building of the Rockies was really a three-stage affair:

• The Colorado Orogeny in Pennsylvanian time, 300 million years ago, created the Ancestral Rockies, two island ranges we call Frontrangia (west of the present Front Range) and Uncompahgria (in southwest Colorado).

• The Laramide Orogeny in early Tertiary time, 65-45 million years ago, established the present ranges of the Rockies.

• Miocene-Pliocene regional uplift 28-5 million years ago raised the whole state and parts of adjacent states to their present elevations.

Between and after these orogenies there were periods of erosion, equally important to shaping today's scenery. After the Colorado Orogeny, the Ancestral Rockies were completely leveled, though debris washed from them survives in the red rocks which surround some of the present ranges. Following Laramide uplift there was a pause, a long stable breather, with little or no upward movement. Erosion played a winning game again, and streams and rivers carved the mountains, carrying rock and gravel, sand and silt to the foot of the mountains, there to spread their loads in wide aprons of debris that gradually covered much of the state under hundreds of feet of sediment. The mountains themselves were half buried by it, until they looked like the scattered ranges and peaks that project from wide sloping plains in the deserts of southern Arizona, New Mexico, Nevada, and West Texas.

Sometime between the Laramide Orogeny and the Miocene-Pliocene uplift, throughout much of the mountain area, molten rock stirring deep within the crust sought its way upward. Enormous volumes of magma erupted from several volcanic centers. Much of it never reached the surface. Some spread into cracks and fissures as dikes, sills, and laccoliths. Much larger masses solidified as stocks and small batholiths. As the magma slowly cooled, crystals grew within it — not the interlocking crystals of the ancient granites but small, isolated mineral crystals in a finer, often glassy groundmass —forming a rock called **porphyry.**

During Miocene and Pliocene time, the whole state, and adjacent states as well, rose into a broad, irregular dome about 5000 feet higher than before uplift, with of course considerable squeezing and deforming, bending and breaking. 5000-foot mountains became 10,000-foot ones, plains only a thousand feet above sea level lifted to 6000 feet, and summits over 9000 feet high became Fourteeners.

Only one slender slice of pie remained low — a long, narrow, crooked sliver represented today by the Rio Grande Valley of New Mexico and the San Luis and Arkansas Valleys of Colorado. Faulted along both edges, this sliver apparently remained almost stationary while the rest of the region was lifted, perhaps because of tensions that developed when the Colorado-New Mexico area was arched upward.

After Miocene-Pliocene uplift, rejuvenated streams and rivers cut into the hundreds of feet of old debris that had covered the mountains, and carried it east and west and south out of Colorado toward the Mississippi Valley, the delta of the Colorado River, and the new-formed Rio Grande Valley of New Mexico. The old Laramide mountains — the Rockies — were unveiled, and began to look as they look today.

Sedimentary layers were stripped from most of the uplifted ridges. The Tertiary pediment — relic of the sloping Miocene plain — was clearly exposed. Developing Pleistocene glaciers chiseled the peaks, widened the mountain canyons, and further shaped the ranges as we know them.

The word "range" is sometimes a puzzler. The great mountain chain of the Rockies stretches from Alaska to New Mexico — some say farther. Within United States it is usually subdivided into Northern, Middle, and Southern Rockies — the

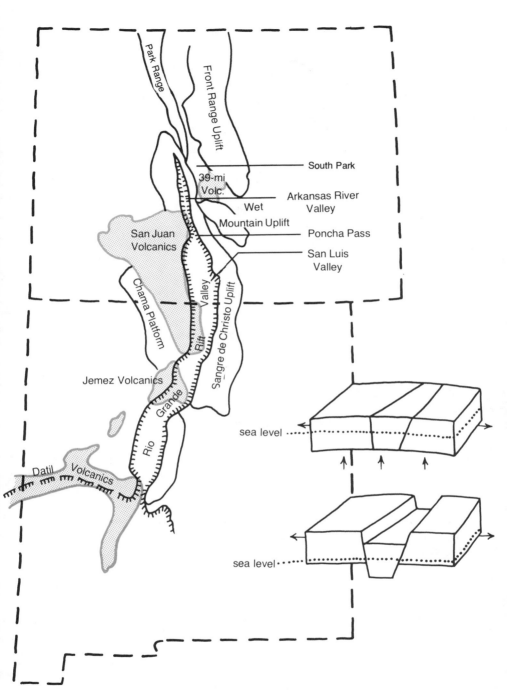

The San Luis Valley and the Arkansas Valley are a northern extension of the Rio Grande Rift, a relatively downdropped block of crust that stretches from the middle of Colorado to the middle of New Mexico. Like rift valleys in other parts of the world, it is associated with areas of volcanism.

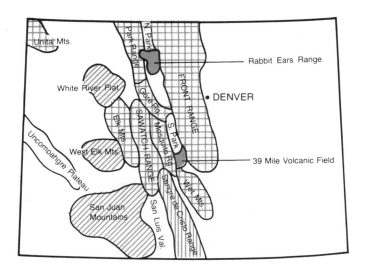

FAULTED ANTICLINES

FAULT BLOCK MOUNTAINS

VOLCANIC MOUNTAINS

DOMES COVERED WITH VOLCANIC ROCKS

Mountain ranges of Colorado

last embracing Colorado's mountains. (To a Canadian, Northern Rockies means Canadian Rockies.)

To confuse the issue, the Colorado Rockies are subdivided into about a dozen individual ranges, shown on the map. They are genuine ranges, individual well defined geologic structures. But each of them is divided again in casual and official usage. Thus the Front Range includes the Mummy Range, Indians Peaks, the Rampart Range, and the Tarryall and Kenosha Mountains. The Collegiate Range is part of the Sawatch Range. The Culebras are part of the Sangre de Cristos. You may find that your road map uses these local names, which **are** useful in a local way. This guide with few exceptions sticks to the names shown on the map, names that refer to geologically meaningful units.

The central cores of most Colorado ranges are composed of hard Precambrian igneous and metamorphic rock – granite, gneiss, and schist. Relatively smooth uplands probably date from pre-uplift times. Pleistocene glaciers carved horseshoe-shaped cirques and deep canyons visible here. South Park, a broad high-altitude intermontane valley, is at the upper left. T.S. LOVERING PHOTO, COURTESY OF USGS

Of Colorado's faulted anticline ranges, the Front Range, though quite a bit larger than the others, is typical. Its core is a long block of hard igneous and metamorphic Precambrian rock: granite, gneiss, and schist 1000 to 1750 million years old. This core, a broken segment of the ancient crystalline rock that underlies the North American continent, rose 15,000 to 25,000 feet during the Laramide Orogeny. As uplift went on, erosion stripped away the sediments that once lay above it; none are now left on the crest of the Front Range. There may never have been a time when sedimentary rocks arched completely across the mountains. Erosion must have begun its attack as soon as uplift started, and sedimentary rocks wash away far more easily than the hard Precambrian mountain core.

The map of the Front Range shows numerous faults edging the mountain block. Many are thrust faults along which up-lifted Precambrian rocks spread outward over the sedimentary rocks. Many geologists believe that the ancient rock, relieved of the pressures that had for a billion years confined it, ex-panded a little, and that the great weight of the rising moun-

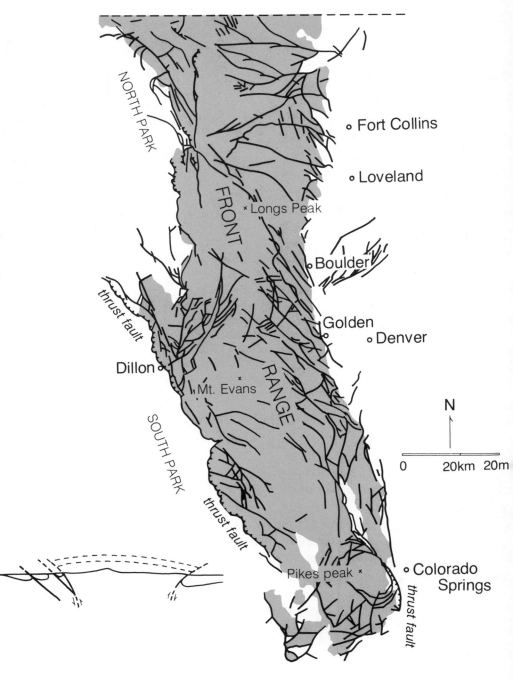

Sketch map showing major faults of the Front Range. All heavy lines are known faults. The Precambrian mountain core is shaded. In thrust faults (⬛ *) Precambrian rocks are pushed out over surrounding sedimentary rocks. Insert is section across range, with no vertical exaggeration, showing hypothetical sedimentary faulted anticline.*

110

tain mass also caused lateral spreading along its flanks. In some places it has ridden out a mile or more over surrounding sedimentary rocks.

The same map also shows the Colorado Mineral Belt as a band of closely spaced faults that crosses the range from Boulder to Dillon. Almost all the mining centers of the Front Range lie along this band. Mineral veins formed during and after the Laramide Orogeny, when mineral-rich solutions seeping upward from somewhere deep within the crust worked their way into fissures and cracks that opened as the mountains rose. In some areas, the solutions accompanied intrusions of igneous magma. Characteristically, mineralized veins — whether or not they contain valuable ores — are stained near the surface by iron oxide in rusty shades of yellow and red. Often this staining led observant prospectors to rich orebodies. Today it still yellows the old mine dumps, and very distinctly marks the limits of the mineral belt.

The Front Range mountain unit is complicated by thousands of less significant faults too small to put on the map. Precambrian rocks of the mountain core are so highly faulted that it is often difficult or impossible even to trace the boundary between igneous and metamorphic rocks! It is equally difficult to trace gold-bearing veins for any distance, as any frustrated miner will tell you.

Faults seem to be particularly abundant in the Colorado Mineral Belt. Do they exist elsewhere in such abundance, yet go unrecognized? Is the local abundance of faults real, and related perhaps to the amount of mineralization? Or do we simply know more about them in mining areas because of the detailed probing which has gone on in search of minerals?

The patterns of erosion in the mountains are varied. Rivers and rivulets carve steep-walled V-shaped canyons. Cloudburst-swollen mountain streams tumble giant boulders like marbles, undercut tall cliffs, and often destroy the works of man. Avalanche, mudflow, and rockslide scar and reshape timbered slopes. And with patient ferocity water seeps into tiny cracks and freezes, its expansion forcing the cracks to widen. Repeated night by night, winter after winter, the process breaks and shatters the solidest rock — freezing-and-thawing is one of the most potent agents of erosion in the mountain area.

Another active erosional agent is chemical weathering, particularly in granite. This type of decomposition occurs mostly because mica crystals in the granite absorb moisture and change chemically into clay minerals. They increase in volume in the process, and, expanding, loosen the quartz and feldspar grains. Wherever granite is exposed, you can quickly recognize it by the rounded knobs and boulders caused by this type of weathering. Day-night and winter-summer temperature changes help the process along.

During the Pleistocene Ice Ages (and even to some degree more recently) glaciers helped to carve the peaks and ridges. Let's review for a moment their causes and effects.

Glaciers are nourished by snowfall. When winter snow exceeds summer melting, snow deepens with each passing year. Gradually recrystallizing, it forms tiny ice granules — you can find them on any summer snowfield — that slowly fuse into solid ice under the weight of added snow. When masses of snow

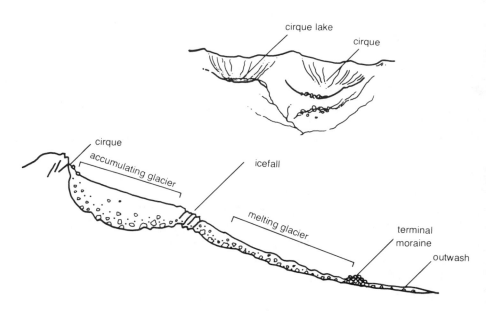

Profile of a valley glacier.

and ice become heavy enough to flow (it usually takes 100 feet or more of packed snow and ice), glaciers are born. Their birthplace may be below high ridges where winter winds whip a lip of snow over the divide from the windward side of the mountains and add it to snow already deposited on the leeward side; it may be on broad, fairly level mountain uplands where deep snow accumulates into fields of ice, and the ice flows slowly in all directions from the highest point. Such upland ice masses are called **icecaps**; in places they finger down into **valley glaciers**.

At least three times glaciers formed in the Colorado Rockies. There may have been earlier glacial episodes, but if there were, evidence of them was erased by one or another of the last three. All three of those recognized here fit into the last of four great glacial epochs known in North America — the Wisconsin glaciation.

As they pluck rock from the heads of valleys or gouge it from the floors and walls and spurs of the valleys, glaciers leave large, clear fingerprints. Now, long after the glaciers have vanished, their semicircular **cirques** mark high mountain ridges and peaks. In them, glaciers literally plucked rocks from the wall by freezing to them and then flowing away, pulling them loose. In valleys, glaciers undermine cliffs as they plow slowly along. Using rocks as cutting tools, the glaciers widen, straighten, and deepen their pathways, giving them U-shaped profiles quite unlike V-shaped stream-cut canyons. Like powerful rasps they groove and striate rocks over which they pass, and like conveyor belts, they dump their loads in **moraines** near their melting lower ends. All these glacial features can still be seen in Colorado's ranges, usually above 8000-foot elevations, for at about 8000 feet the glacier tongues melted faster than they were resupplied from above. Below that level, traces of glaciers are found only in the broad stepped **terraces** deposited by overloaded streams fed by melting ice.

The tiny glaciers existing today in the Front Range, and the glacier-like year-round icefields of the Park Range, are probably not simple remnants of Ice Age glaciers. Geologists now think they may be new bodies of ice formed within the last few thousand years, after a long warming period. Are we entering another Ice Age? Who can tell?

Colorado's water resources are greatest on the western side of the mountains — an accident of prevailing westerly winds. Her water needs are greatest in the band of cities east of the mountains. As you drive through the mountains, you will often see reservoirs, water conduit lines, and tunnel exits that are part of the huge network of water projects that increase eastern slope supplies.

Breckenridge, once a bustling gold camp, now attracts skiers and summer tourists. Ski runs course across a moraine below two cirques in the Tenmile Range. Timberline is at about 11,000 feet here.

JACK RATHBONE PHOTO

interstate 70 (and u.s. 6)
denver — dillon

(66 miles)

At the geologic site at the I-70 roadcut through the Dakota Hog-back, you can see a cross-section of Mesozoic rocks tilted steeply against the mountain's flank by uplift of the Front Range. Just east of the roadcut the highway crosses the Golden Fault Zone, the main line of faulting that edges the east side of the Front Range. But the fault is hidden, as most large faults are, by soft rock weathered into soil. West of the fault, the rock layers are intact down to the Precambrian rocks that are the core of the range.

Gneiss and schist are the first Precambrian rocks you'll see as you enter the mountains. These ancient metamorphic rocks, 1750 million years old, form the core of the range here, and the highway carves through them for about 25 miles. They show obvious signs of the pushing and wrenching, melting and squeezing, to which they have been subjected during repeated periods of mountain building. In many places (as at Mile 257) they are crisscrossed by light-colored pegmatite dikes with oversized quartz and mica crystals.

Near Idaho Springs and Georgetown, where the highway crosses the Colorado Mineral Belt, numerous dikes, or veins as miners call them, contain ores of gold and other metals. Many of these have been dug or blasted completely away, for this was the principal mining

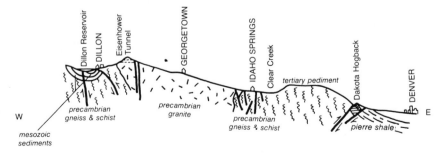

Section along I-70 Denver to Dillon. Arrows show direction of movement.

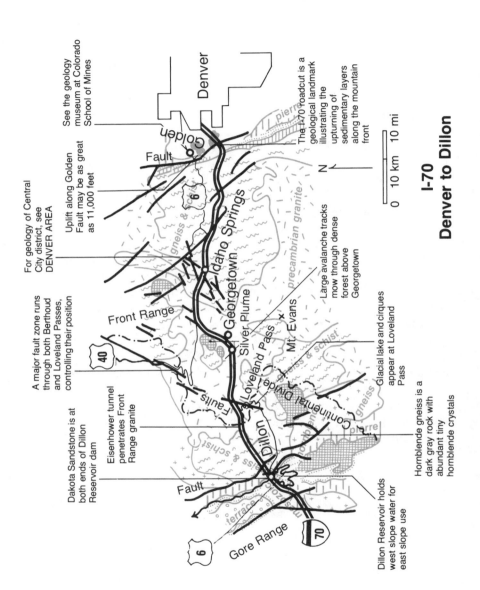

I-70
Denver to Dillon

0 10 km 10 mi

N

See the geology museum at Colorado School of Mines

The I-70 roadcut is a geological landmark illustrating the upturning of sedimentary layers along the mountain front

Uplift along Golden Fault may be as great as 11,000 feet

For geology of Central City district, see DENVER AREA

Large avalanche tracks mow through dense forest above Georgetown

A major fault zone runs through both Berthoud and Loveland Passes, controlling their position

Glacial lake and cirques appear at Loveland Pass

Dakota Sandstone is at both ends of Dillon Reservoir dam

Eisenhower tunnel penetrates Front Range granite

Hornblende gneiss is a dark gray rock with abundant tiny hornblende crystals

Dillon Reservoir holds west slope water for east slope use

Denver

Golden

Fault

Idaho Springs

Georgetown

Silver Plume

Loveland Pass

Mt. Evans

Continental Divide

Front Range

Dillon

Fault

Gore Range

precambrian granite

gneiss & schist

pierre

gneiss & schist

hornblende gneiss & schist

terrace gravels

Resistant Dakota Sandstone at the east end of the I-70 highway cut, with its distinctive black coaly layers, normally protects the soft purple and green and brown shales in the Morrison Formation at the west end. Notice that the Morrison Formation is usually soil-covered where it has not been sliced open by highway construction.

district in Colorado from 1860 to the late 1880s. More than $400 million worth of gold, lead, zinc, and copper have been produced from this area.

A few miles west of Idaho Springs, I-70 gets into glaciated country above the 8000-foot contour. The first hilly, well defined terminal moraine is near the junction with U.S. 40. Above it, the valley assumes the U-shaped profile that is the hallmark of valley glaciation, though rock slides, avalanche debris, and dense forest sometimes obscure it.

At Georgetown, a new covey of condominiums is built on alluvial fans on the south side of the valley. Their owners live dangerously in the direct path of flood and avalanche. Numerous avalanche tracks mark the slopes above them, and those big boulders between the buildings once came crashing down the mountainside!

Patches of granite begin to appear near Georgetown. This is the Silver Plume Granite, 1450 million years old, named for the next town upstream and a waterfall destroyed by highway construction. You will still see irregular bands of metamorphic rocks, however, as the granite has pried its way in here in slender fingers branching from the edges of a batholith.

117

The wide avalanche track at Mile 218 has been a headache to highway maintenance crews for decades. The ridge-like berm parallel to the highway was built to deflect and break up future snowslides in an effort to protect the highway.

As the Interstate approaches Loveland Basin, either go straight ahead through the Eisenhower Tunnel or follow U.S. 6 over scenic Loveland Pass. Along either route the road is bordered by alternating patches of gray granite like that at Silver Plume, and gneiss and schist like that at Idaho Springs. Along the Loveland Pass road, good outcrops are rare because the rock is badly shattered near a broad fault zone that controls the position of the pass. The pass developed where erosional processes exploited a band of broken, weakened rock along the fault zone. The same zone extends northeast to Berthoud Pass and beyond. Loveland Pass is on the winding Continental Divide, backbone of the continent, and the Eisenhower Tunnel drills through the mountain below the divide. Here streams flow east to the Atlantic, or west ultimately to the Pacific, and here Colorado's eastern slope meets the western slope.

Loveland Pass looks south to three cirques, arcuate bites incised by glaciers, that frame Arapaho Basin ski area. To the northwest, a larger cirque curves around Loveland Basin. The smoothly contoured uplands here were rounded by Pleistocene ice that covered them completely, but Pettingell Peak to the north, contrastingly sharp and craggy, jutted up through the icecap.

Pass Lake just south of the pass occupies another glacial cirque. Huge talus cones form where rock loosened from the cliffs by freezing and thawing tumbles down ravines toward the lake.

The Idaho Springs Formation displays the alternating bands of light granular crystalline minerals (mostly quartz and feldspar) and dark fine-grained or platy minerals (mostly biotite and hornblende) that characterize gneiss. T.S. LOVERING PHOTO, COURTESY OF USGS

118

Breaking along joints and schist surfaces, and probably loosened by freezing and thawing, boulders of Precambrian gneiss crashed onto Interstate 70 in May, 1973. The largest of them weighed 200 tons.

W.R. HANSEN PHOTO, COURTESY OF USGS

Both I-70 and U.S. 6 descend toward Dillon through glaciated valleys, but many glacial features along I-70 are hidden beneath landslide debris. The slides pose serious problems for highway engineers. Near Mile 211 you can see how they have been handled where the toes of two large slides reach the highway. Broken and tilted trees show that despite engineering efforts the slides are still moving.

Along U.S. 6 several terminal moraines can be seen, less by the shape of the moraine as a whole than by the appearance of jumbled unsorted rock and sand and silt in highway cuts. The hummocky, irregular moraines are hard to discern because they usually support a dense forest of pines and Douglas fir, trees that like a loose, well drained foothold.

Along both highways Cretaceous rocks appear east of Dillon Reservoir Dam — steeply west-dipping Pierre Shale and Dakota Sandstone. Sometimes they are overturned and seem to dip east. They are overridden from the east by Precambrian metamorphic rocks (at Mile 208 on I-70). Although the fault along this side of the mountains may be vertical deep below the surface, near the surface it arches westward as the crystalline rocks expand and spread out.

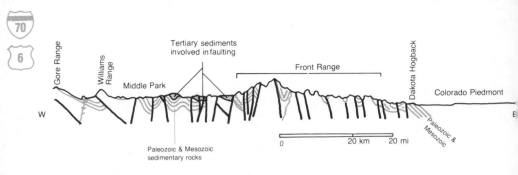

During the boring of the many tunnels that now bring western slope water to the eastern slope, geology within the Front Range was found to be much more complex than had been suspected.

Dillon Reservoir stores western-slope water for the City of Denver. Water collected here is pumped 23 miles through a tunnel under the Continental Divide, a tunnel carefully studied by geologists for its glimpse into the interior of the Front Range. Although there is considerable resistance to further "theft" of western-slope water, the needs of the eastern-slope cities have increased our opportunities for water sports and lakeside recreation, as well as our knowledge of the complex Precambrian geology inside the Front Range.

interstate 70 (and u.s. 6)
dillon — dotsero
(76 miles)

Below Dillon Reservoir Dam, I-70 crosses the Blue River, a Colorado tributary, as it flows northwest along the faulted syncline between the Front Range and the Gore Range. Curving across a broad boulder-strewn moraine, the highway abruptly enters Tenmile Canyon between the Gore and Tenmile Ranges. In terms of geologic structure, these two ranges are continuous; faults and Precambrian-Paleozoic contacts along their boundaries can be traced right across Tenmile Canyon.

This narrow defile opens the heart of these ranges to view, for it gouges deeply into their Precambrian core along the line of a particularly large fault. The smooth rock surface on the east wall of the canyon, to your left beyond Mile 196, is an actual fault surface. Shiny areas visible when the sun happens to be at the right angle are **slickensides** produced by grinding and rasping and sliding of rock against rock. On that wall the rock is dark gray hornblende gneiss with only a few pegmatite veins. On the west side of the canyon, west of the fault, Precambrian rocks are beautifully exposed in both natural outcrops and huge new highway cuts such as the one at Mile 199. The strongly banded black and white gneiss, with contorted flow lines, is cut by dozens of large veins and literally hundreds of little veinlets.

The highway turns west at Copper Mountain and leaves the fault, which goes on up Tenmile Canyon. Just opposite Copper Mountain the road passes from the striped gneiss to reddish brown granite. Then, hummocky reddish glacial deposits and landslide debris cover all the older rocks and hide the great thrust fault along which the granite pushes westward over younger rocks.

The younger rocks, however, soon appear at the surface. Do they look familiar? Red sediments for all the world like those at Red Rocks near Denver, deposited along the west side, rather than the east side, of Frontrangia. They, like their eastern counterparts, are dull red sandstone and conglomerate interlayered with dark red shale. They

121

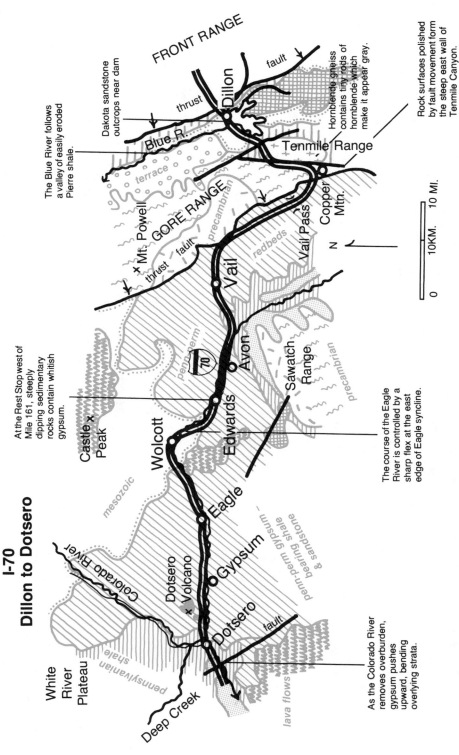

I-70
Dillon to Dotsero

FRONT RANGE

fault

Dillon

thrust

Dakota sandstone outcrops near dam

Blue R.

Hornblende gneiss contains tiny rods of hornblende which make it appear gray.

Tenmile Range

Rock surfaces polished by fault movement form the steep east wall of Tenmile Canyon.

The Blue River follows a valley of easily eroded Pierre shale.

terrace

Mt. Powell

GORE RANGE

precambrian

Copper Mtn.

Vail Pass

thrust fault

Vail

redbeds

N

0   10KM.   10 MI.

Sawatch Range

precambrian

Avon

penn-perm

70

At the Rest Stop west of Mile 161, steeply dipping sedimentary rocks contain whitish gypsum.

Edwards

The course of the Eagle River is controlled by a sharp flex at the east edge of Eagle syncline.

Castle Peak x

Wolcott

mesozoic

Eagle

penn-perm gypsum - bearing sandstone & shale

Colorado River

Dotsero Volcano x

Gypsum

White River Plateau

pennsylvanian shale

Dotsero

fault

Deep Creek

As the Colorado River removes overburden, gypsum pushes upward, bending overlying strata.

lava flows

122

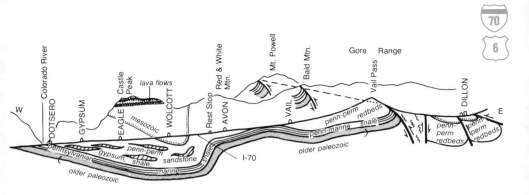

Section along I-70 from Dillon to Dotsero

are thousands of feet thick here and the highway remains in them for nearly 18 miles. Their exact age is a mystery; they contain no fossils by which they can be dated. But they are called Pennsylvanian-Permian because they are sandwiched between Lower Pennsylvanian and Triassic rocks.

The red rocks are underlain by Pennsylvanian brown and gray sandstone and shale, well exposed between Miles 172 and 193. The shale layers are weak, and tend to move easily, especially when they are soaking wet. So they offer new challenges to highway engineers. The bumps in I-70 near Vail are due to the tendency of these rocks to slump and slide. They also hang dangerously steeply above the Vail ski resort.

Between Vail and Dotsero, the highway crosses first an anticline and then a broad syncline in these sediments. Rocks at Vail rise westward, flex sharply downward along the Eagle River between Edwards and Wolcott, and then rise again toward Dotsero. Crossing these flexes you will also see a gradual change in the rocks themselves, for the redbeds give way westward to light gray shale containing a lot of gypsum, and eventually to true marine dark gray shales. There are many minor flexes and small faults to make interpretation difficult, but watch for changes in dip from westward-dipping rocks at Vail Pass to eastward-dipping ones at Vail, and then westward-dipping ones again near Edwards.

At Avon, and again at Gypsum, bluffs along the Eagle River contain sediments rich in gypsum. White stringers and occasional large white gypsum blobs can be seen fairly easily from the highway. Gypsum forms when sea water dries up. Fairly common on a worldwide scale, it occurs here in a little area centering in the valley of the Eagle River, where it accumulated more than 280 million years

123

ago as the narrow sea between the two ranges of the Ancestral Rockies evaporated.

Near Mile 157, Pennsylvanian sedimentary rocks give way to younger ones. They are best seen after you pass the deepest part — the **axis** — of the syncline. Watch for:

• Cretaceous limestone forming a yellowish surface on hills across the river at Mile 157.

• Black Benton Shale on hill slopes below the limestone.

• Thick, resistant Dakota Sandstone, with coaly beds, capping hills on both sides of the river at Mile 156.

• Varicolored purple and green shale of the Jurassic Morrison Formation, across the river at Mile 156.

• Blocky peach-colored Jurassic sandstone at Mile 154.

• Bright orange-red Triassic siltstone at Mile 153.

• Pennsylvanian gypsum-bearing shales again. Here they are more than 1000 feet thick. In bluffs along the Eagle River between the towns of Eagle and Gypsum, they are as contorted as marble cake, with gray and white and brown swirls and folds that do not seem to relate to parallel-bedded sedimentary rocks elsewhere. Their contortions show particularly well on the north bank of the river east and west of the town of Gypsum. Gypsum — the mineral — can flow in a solid state, given time, in much the same manner as glacial ice or silly putty, and here is good evidence of this type of flow. Notice that weathering of the gypsum-rich shale produces monotonously poor soil on which few plants can grow.

Near the town of Gypsum, Pennsylvanian sediments are folded and contorted because of the upward flow of gypsum layers where overlying rock layers have been removed by the river.

T.S. LOVERING PHOTO, COURTESY OF USGS

Just west of Mile 136, near Dotsero, the black rock in sagebrush north and south of the highway is a basalt lava flow which comes from a small gulch in the gypsum bluffs north of the highway. Tucked into the hills is a little volcano, mostly a cinder cone but with the basalt flow coming from a vent near its base. At present the volcano is being quarried for its foamy, lightweight volcanic cinder, used for concrete blocks. Volcanism of this type is quite unusual in Colorado, and since this is also Colorado's youngest volcano, a little over 4000 years old, interested geologists are campaigning for its rescue and preservation.

Where the Eagle River joins the Colorado near Dotsero, Pennsylvanian rocks are gray and brownish gray marine shales, quite different from either the red sandstone and conglomerate close to ancient Frontrangia or the gypsum-bearing sediments near Wolcott. These rocks occasionally contain thin beds of low-grade coal, indicating that densely vegetated swamps edged the embayment in which they were deposited. There is still quite a bit of gypsum in lower layers, and along the Colorado River west of the bridge the gypsum has pushed upward toward the line of the river canyon, steeply bowing up the dark shale. If you press hard with two fingers on a piece of silly putty, you can see what has happened. The putty tends to squeeze up between your fingers, where there is less pressure. That's what the gypsum has done here, where the river has removed some of the weight and pressure of overlying rocks.

At Dotsero, the road to the right just west of the bridge follows the Colorado River upstream for about 50 miles, and though unpaved makes an interesting and informative side trip. The river cuts down through Pennsylvanian and Mesozoic sedimentary rocks and reveals the complex nature of the syncline between the Gore Range and the White River Plateau to the west. Two miles up this road is Deep Creek, and 1½ miles up Deep Creek (by dirt road and trail) there are lots of Devonian and Mississippian fossils. The rock sequence is the same as that given in the next section.

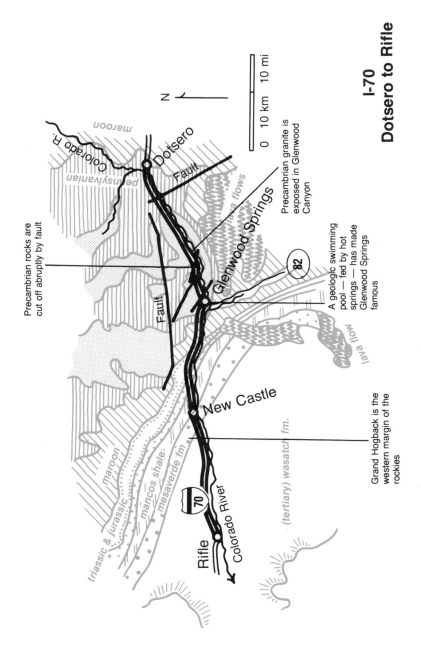

Precambrian rocks are
cut off abruptly by fault

maroon

Colorado R.

Dotsero

Fault

pennsylvanian

Precambrian granite is
exposed in Glenwood
Canyon

Glenwood Springs

lava flows

Fault

82

A geologic swimming
pool — fed by hot
springs — has made
Glenwood Springs
famous

lava flow

New Castle

maroon

triassic & jurassic

mancos shale

mesaverde fm.

Grand Hogback is the
western margin of the
rockies

70

Colorado River

Rifle

(tertiary) wasatch fm.

N

0   10 km   10 mi

I-70
Dotsero to Rifle

Cambrian, Devonian, and Mississippian sedimentary rocks lie on coarse pink Precambrian granite in Glenwood Canyon. The beveled surface of the granite is just above the railroad tunnel.
JACK RATHBONE PHOTO

interstate 70 (and u.s. 6)
dotsero — rifle

(43 miles)

West of Dotsero, Interstate 70 (U.S. 6) enters Glenwood Canyon, a 15-mile-long gorge excavated by the Colorado River. Most of the carving of the gorge was done in Pleistocene time, when run-off was far greater than now and floods were the usual state of affairs. But you can't see the muddy river in spring without realizing that the process is still going on. The location of the river was governed originally by the contact between hard Mississippian limestone and soft Pennsylvanian shale, but now the canyon has cut down through other Paleozoic layers into Precambrian granite, following vertical joints in the ancient rock.

The Paleozoic rocks are fairly easily recognized in passing. In the order you will come to them, they are:

• Mississippian limestone, a massive cliff-forming gray limestone about 260 feet thick, visible high up to the west as you enter the canyon (Mile 131). This is the Leadville Limestone.

• Devonian thin-bedded limestone and shale forming greenish gray slopes and ledges below the massive Mississippian.

• Ordovician dolomite (like limestone but with high magnesium content) and shale that form brown-tinted cliffs near the highway at the Garfield County line.

• Cambrian quartzite (silica-cemented sandstone) recognizable by its light and dark brown banding. A 400-600 foot cliff-former (Mile 130), this formation lies right on:

• Coarse pink Precambrian granite, which comes to the surface at Mile 128.

At Mile 128, a fault brings Pennsylvanian rocks to roadside level again, and the sequence is repeated.

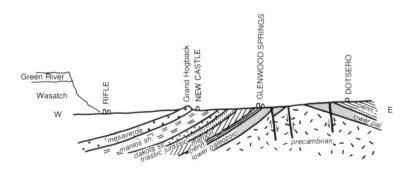

Section along I-70 Dotsero to Rifle

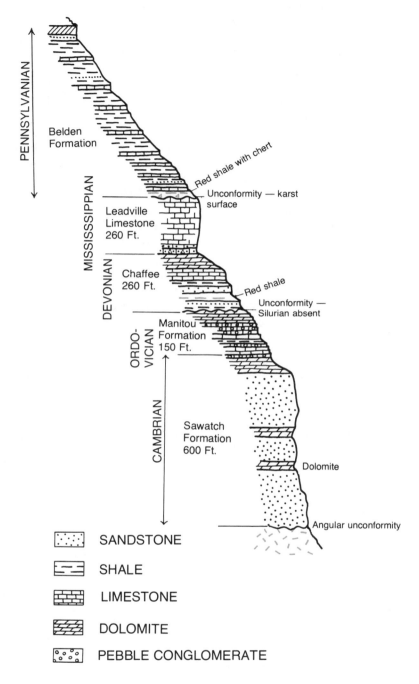

PENNSYLVANIAN

Belden
Formation

MISSISSSIPPIAN

Red shale with chert

Unconformity — karst
surface

Leadville
Limestone
260 Ft.

DEVONIAN

Chaffee
260 Ft.

Red shale

Unconformity —
Silurian absent

ORDO-
VICIAN

Manitou
Formation
150 Ft.

CAMBRIAN

Sawatch
Formation
600 Ft.

Dolomite

Angular unconformity

SANDSTONE

SHALE

LIMESTONE

DOLOMITE

PEBBLE CONGLOMERATE

Stratigraphic column of rocks exposed in Glenwood Canyon

129

Shoshone Dam diverts Colorado River water into a tunnel leading to a power plant two miles below. Near the dam and power plant gray granite, garlanded with attractive pink and white veins, forms the canyon walls. Its contact with Cambrian sandstone is abrupt, and represents a long gap in the record, the Great Unconformity spanning the half billion years of the Lipalian Interval. The featureless surface of the Precambrian illustrates the featureless landscape that forms during such a long period of erosion.

West of the power plant, Paleozoic sediments — dipping west now — come down to the level of the highway, and you progress from older to younger rocks, the reverse of the progression outlined above. A fault zone at Mile 119 abruptly cuts off the downstream end of Mississippian rocks and brings up Precambrian granite again. The highway tunnel is in this granite. West of the tunnel the older part of the section is repeated yet again, with the Precambrian-Cambrian contact at Mile 118 east of Glenwood Springs.

After retreat of the sea in which the Leadville Limestone was deposited, and before Pennsylvanian sedimentation, a **karst** surface developed on the Mississippian limestone. Watch the top of the gray limestone cliff near Miles 119 and 120 for solution caves, sink holes, and reddish soil full of broken limestone blocks. This type of surface, not uncommon over limestone layers, exists today in Yugoslavia, Puerto Rico, large areas of southeastern United States, and many other regions. Small modern caves show here, too.

Hotsprings at Glenwood flow from Pennsylvanian shales like those at Dotsero. The hot water rises rapidly and probably from considerable depth through broken and shattered rocks in a fault zone along the river. The springs, right by the river at the east edge of town, are difficult to see from the highway; the big hotsprings pool at Glenwood is fed by them. Total flow of all the springs approaches 2500 gallons per minute, and the water temperature is 123°F.

The massive gray rock north of I-70 just west of Glenwood Springs is Mississippian Leadville Limestone again. West of it are more contorted gypsum-bearing Pennsylvanian shales and then the Maroon Formation, another equivalent of the red rocks formed along the flanks of Frontrangia in Pennsylvanian and Permian time. These sediments, more than 3000 feet of them, were washed off the western highlands of Uncompahgria. Above them lie, successively, bright red Triassic shale, massive light brown Jurassic sandstone, and our old friend the varicolored, slope-forming Morrison Formation. Finally, the Cretaceous Dakota Sandstone, also an old friend if you are driving from the east.

Farther down the valley, about nine miles from Glenwood, gray Cretaceous shale forms the valley floor. This is the Mancos Shale, above the Dakota Sandstone. It takes the place of the Pierre Shale on this side of the mountains, but is actually a little older than the Pierre.

Above the gentle Mancos slope rise cliffs of the Mesaverde Group, a series of thick, resistant, near-shore Cretaceous sandstones with many coal layers and carbonaceous shales.

As the highway cuts through the Dakota Sandstone, it swings northward along the Mancos Shale valley between the Dakota Hogback and a much larger, much more prominent ridge called the Grand Hogback, formed by the Mesaverde Group.

At New Castle, the rivercut through the Grand Hogback almost matches the manmade cut through the Dakota Hogback near Denver. Several old coalmine tunnels and dumps mark the side of the hogback. In northwestern Colorado, Wyoming, and Montana, coal seams of this same age may be 50 feet thick. They are one of the greatest energy reserves in this country. Red streaks on slopes below the Mesaverde cliff mark burned shale, or clinker, baked like pottery clay by burning coal seams.

Not far west of New Castle the highway passes out of Cretaceous rocks and into Tertiary ones. The narrow gap just west of New Castle goes through the youngest of the Cretaceous rocks, a shoreline sand and conglomerate deposited as the sea drew back eastward from the region. Coarsening debris in this series of beds suggests that mountains were rising as the sea retreated.

Like the Dakota Hogback east of the mountains, the Grand Hogback outlines the Colorado Rockies for many miles. Once west of it you are officially out of the Rocky Mountains and into the plateau country.

JACK RATHBONE PHOTO

Most of the Tertiary rocks here belong to the Wasatch and Green River Formations, which form high plateaus sometimes capped by younger lava flows. The Wasatch is mostly sandstone and shale of a monotonous mouse-gray color, with tinges of pink, the silty shale that gives the town of Silt its name. High up on the cliffs to the north, in the Green River Formation, are dark brown ledges of oil shale, another energy reserve, discussed in I-70 RIFLE — GRAND JUNCTION.

Notice particularly how poorly consolidated the Tertiary rocks are. They are not nearly as durable as Paleozoic ones farther east. Volcanic rocks are an exception, for lava flows are hard as soon as they solidify from molten magma.

From Rifle, an interesting side trip on Colorado 13/789 and Colorado 325 leads to Rifle Gap. This natural break in the Grand Hogback attained fame a few years ago as the site for a controversial but short-lived orange nylon "sculpture." Behind the Gap is Rifle Box Canyon, a narrow oasis carved in Mississippian Leadville Limestone, and some remarkable travertine terraces. The itinerary for CO. 13/789 CRAIG — RIFLE gives a description of these features.

u.s. 24

colorado springs — buena vista

(94 miles)

Plunging into the mountains in a northwest direction, U.S. 24 follows very closely the line of the Ute Pass Fault, one of the major faults on the east side of the Front Range. The fault slices through the mountains, separating the Rampart Range from the rest of the Front Range. This fault has been traced for about 60 miles, from well north of Woodland Park to the south end of Cheyenne Mountain. The eastern elevated side was raised more than 1000 feet higher than the western side. Now both sides are beveled, except at Pikes Peak and a few other high points, by the Tertiary pediment, the erosion surface that most geologists regard as having once been continuous with the High Plains.

The fault zone is quite wide, and broken and fractured Precambrian and Paleozoic rocks along it are shattered enough to erode

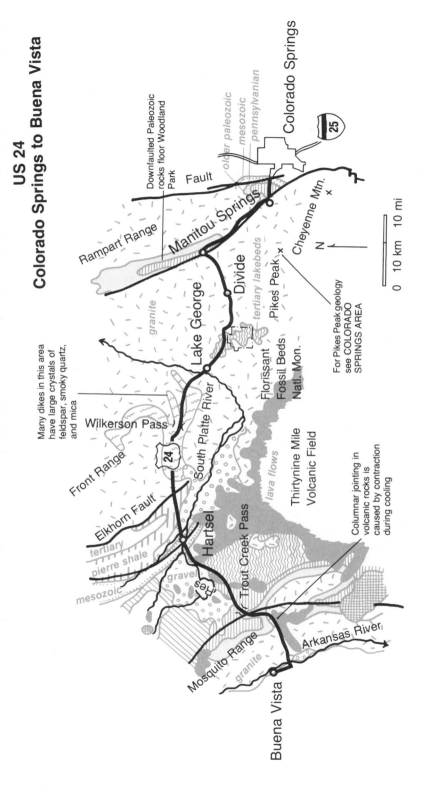

US 24
Colorado Springs to Buena Vista

Downfaulted Paleozoic rocks floor Woodland Park

Fault

older paleozoic
mesozoic
pennsylvanian

Colorado Springs

Rampart Range

Manitou Springs

Cheyenne Mtn. ×

granite

Divide

Lake George

tertiary lakebeds

Pikes Peak ×

N

0 10 km 10 mi

For Pikes Peak geology see COLORADO SPRINGS AREA

Many dikes in this area have large crystals of feldspar, smoky quartz, and mica

Wilkerson Pass

South Platte River

Front Range

Florissant Fossil Beds Natl. Mon.

Thirtynine Mile Volcanic Field

lava flows

Columnar jointing in volcanic rocks is caused by contraction during cooling

24

Elkhorn Fault

tertiary

pierre shale

mesozoic

gravel

Res.

Hartsel

Trout Creek Pass

Mosquito Range

granite

Arkansas River

Buena Vista

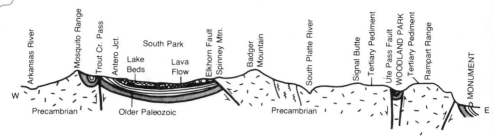

Beeline section parallel to U.S. 24 from Colorado Springs to Buena Vista

easily. Near both Manitou and Woodland Park, Paleozoic rocks still lie in place on the Precambrian granite of the Rampart Range. Near Manitou, they are represented by Cambrian sandstone and Ordovician, Devonian, and Mississippian limestone, well exposed in Williams Canyon (see CAVE OF THE WINDS), and coarse red sandstone and conglomerate of the Pennsylvanian Fountain Formation, in the western part of Garden of the Gods. At Woodland Park these rocks form a long narrow syncline trending northwest along the fault.

Rampart Range, northeast of the fault, is made of the same Pikes Peak Granite as the area southwest of it, an ancient batholith intruded as molten rock or magma about a billion years ago. Typically it is a pink granite, with chunky crystals of pink feldspar, glassy quartz, and black mica or biotite. In many places it is cut by pegmatite dikes, which show up as bands of larger crystals. Elsewhere it may be cut by white veins composed of light-colored mica (muscovite) and white milky quartz.

Half a mile west of Woodland Park, the highway turns southwest and crosses Ute Pass Fault, where the sheared and broken granite along the fault has weathered exceptionally deeply.

By Mile 280 the highway is on the 8000- to 10,000-foot Tertiary pediment, which lines up with the same level on the Rampart Range. Notice the deep weathering of the granite here; it dates from Tertiary time when the pediment was formed. In some places along U.S. 24 the granite has turned gradually, without moving, into a boulder conglomerate or a deep coarse sand. Glimpses of Pikes Peak, southeast of you now, show broad, smoothly contoured, glacially scoured uplands. Products of weathering common at lower elevations have been removed by glacial erosion and wind.

Near the town of Florissant, roadside rocks change suddenly to thinly bedded, fine-grained light gray shale. These are the Florissant Lake Beds, formed in Oligocene time when volcanic ash showered into a mountain lake. The ash falls brought down many insects, leaves, and even birds, now preserved as fossils in the thin gray beds (see FLORISSANT FOSSIL BEDS NATIONAL MONUMENT.)

At Mile 268 near Lake George, Crystal Peak can be seen in the distance to the north. Large feldspar and smoky quartz crystals of museum quality abound in the pegmatite veins in this hill. West of Lake George, as well as near Wilkerson Pass, quartz and pegmatite dikes are well exposed nearer the highway, though the crystals are of poorer quality.

As U.S.24 approaches Wilkerson Pass, metamorphic rocks take the place of Pikes Peak Granite. About 1750 million years old, they often look as though large granite blobs were injected into them, as at Mile 255-256. And so they were; the borderline between granite and gneiss is not at all well defined. Younger granite (1350 million years old), intruding in a molten state, partly melted its way between bands of gneiss and schist. Right at Wilkerson Pass the Precambrian rocks are concealed by Tertiary volcanic flows from the Thirtynine Mile Mountain volcanic field to the southwest. A small Forest Service information center at the pass explains some of the geology of the region.

From Wilkerson Pass look westward over South Park, one of Colorado's four big intermountain valleys, to the Mosquito Range, and beyond it to the high peaks of the Sawatch Range. The Thirtynine Mile volcanic field, possibly the source of volcanic ash at the Florissant Lake Beds, closes the valley on the south. South Park is a lopsided faulted syncline, its east side bordered by the Elkhorn Fault, in which Precambrian gneiss and granite thrust westward over Cretaceous and early Tertiary rocks of the valley floor. On the west, South Park is edged by faults and the tilted sedimentary rocks of the Mosquito Range.

From the center of South Park can be seen the dark double summits of Buffalo Peaks. These peaks contain many thick layers of lava and volcanic ash composed chiefly of feldspar and dark minerals, without much quartz. Near Hartsel, a ridge of tilted Mesozoic sedimentary rock — basically similar to the Dakota Hogback east of the mountains — bisects the Park.

You may wonder about the white deposit on flat parts of the floor of South Park. The park was once called Valle Salada or Bayou Salado because of this material. The flat surfaces are old lake deposits, and the white coating was deposited as the lakes, which were saline, evaporated. It is a mixture of salts, including common table salt — sodium chloride or halite — and a mineral closely resembling Epsom salts. Taste it if you doubt this.

Beyond Antero Junction, U.S. 24 and U.S. 285 follow the same route over Trout Creek Pass. For the geology here see U.S. 285 FAIRPLAY — PONCHA SPRINGS.

Like fossils of leaves and other insects, this bee is preserved as a fine film of carbon darkening the light-colored volcanic ash beds at Florissant National Monument

U.S. NATIONAL PARK SERVICE PHOTO

florissant fossil beds
national monument

(5½-mile side trip from u.s. 24)

At Florissant, fossil leaves, insects, fish, and even a marsupial mammal have been exquisitely preserved in fine layers of volcanic ash laid down in ancient Lake Florissant. Here, nearly 35 million years ago, a lake formed when flows of lava dammed a mountain valley. Curved like a crescent, the lake was 12 miles long and two miles wide. It was swampy in places, and must have attracted an abundance of plant and insect life. Graceful palms and towering sequoias growing by its banks harbored birds and small animals almost like those we know today.

This petrified Sequoia stump is one of several fossil tree stumps at Florissant Fossil Beds National Monument. An apparently luxuriant grove was buried in volcanic ash bout 35 million years ago, and groundwater deposited silica in the cells of the original wood.
U.S. NATIONAL PARK SERVICT PHOTO

During the time when the lake existed, volcanoes occasionally filled the air with volcanic ash. Carried on the wind, some of the ash fell on Lake Florissant, bringing down thousands of insects, killing fish, and burying drifting leaves and bits of wood. The ash settled gently and quickly and prevented rapid decomposition. Other ash, deposited in thick unstable layers on surrounding hills, washed into the lake and valley in mudflows that buried trees and small animals. Leaves and wood, insects and fish and other animals became fossilized under nearly ideal conditions, with even the delicate tracery of wings and leaves preserved.

Rainwater and groundwater filtering through the ash deposited silica in the cells of both wood and bone, petrifying and preserving them. Like an early Pompeii, the lake and the record it held may then have been deeply buried by more volcanic eruptions.

137

Much later, less than 28 million years ago, another geological accident brought Lake Florissant's deposits back to the surface. Miocene-Pliocene uplift of the Southern Rockies led to increased erosion that exposed the old lake sediments. Since their discovery in 1874 by a U. S. Geological Survey geologist, the lake beds have yielded thousands of museum and study specimens — more than 1100 species of insects such as dragonflies, butterflies, ants, flies, beetles, and bees; fossil leaves from birch, willow, maple, beech, and hickory; needles from fir trees and fronds of giant sequoias. Petrified tree stumps show that some of these trees were every bit as large as the great redwoods growing today in California. The fossils indicate that a warm, almost a tropical climate graced this area in middle Tertiary time, when it was at a much lower, warmer elevation.

High peaks of the Collegiate Range, part of the Sawatch Range, rise beyond Buena Vista. Glaciers shaped the high country but did not extend far onto the Arkansas Valley floor. JACK RATHBONE PHOTO

u.s. 24

buena vista — dowd

(72 miles)

Between Buena Vista and Leadville, U.S. 24 follows the Arkansas River Valley between the Mosquito Range and the Sawatch Range. From south to north, visible Sawatch Range Fourteeners are: Mt. Princeton (southwest of Buena Vista), Mt. Yale, Mt. Columbia, Mt. Harvard, and Mt. Oxford, making up the Collegiate Range. North of

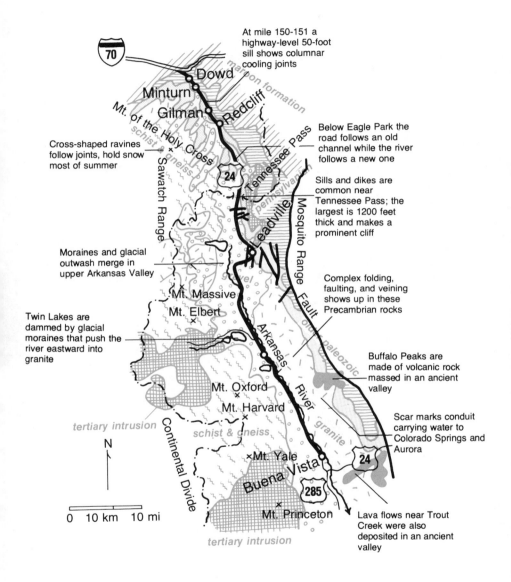

US 24
Buena Vista to Dowd

At mile 150-151 a highway-level 50-foot sill shows columnar cooling joints

Cross-shaped ravines follow joints, hold snow most of summer

Moraines and glacial outwash merge in upper Arkansas Valley

Twin Lakes are dammed by glacial moraines that push the river eastward into granite

Below Eagle Park the road follows an old channel while the river follows a new one

Sills and dikes are common near Tennessee Pass; the largest is 1200 feet thick and makes a prominent cliff

Complex folding, faulting, and veining shows up in these Precambrian rocks

Buffalo Peaks are made of volcanic rock massed in an ancient valley

Scar marks conduit carrying water to Colorado Springs and Aurora

Lava flows near Trout Creek were also deposited in an ancient valley

70

Dowd

Minturn

Gilman

Redcliff

maroon formation

Mt. of the Holy Cross

schist & gneiss

Sawatch Range

Tennessee Pass

pennsylvanian

Leadville

24

Mosquito Range Fault

devel

ordovician paleozoic

×Mt. Massive

Mt. Elbert ×

Arkansas River

granite

Mt. Oxford ×

Mt. Harvard ×

tertiary intrusion

schist & gneiss

N

Continental Divide

×Mt. Yale

Buena Vista

285

×Mt. Princeton

24

0 10 km 10 mi

tertiary intrusion

139

Lake Creek rise Mt. Elbert (14,431 feet, highest in Colorado) and Mt. Massive. The big alluvial fans below these peaks were built by torrential Ice-Age streams that spread heavy burdens of rock and sand as they reached the valley floor. Cirques high on mountain slopes, U-shaped valleys between peaks, and glacial moraines that block lower ends of valleys are further manifestation of glaciation.

Most of the glaciers advanced no farther than the mountain edge. The Arkansas Valley itself does not seem to have been glaciated this far south, although individual ice tongues from the side valleys sometimes reached into or even across it. The valley contains several hundred feet of stream and lake deposits, however. You can see cross sections of some of these deposits in roadcuts, and farther north you can tell which are glacial and which are stream-deposited, for stream deposits are layered (stratified) collections of smooth, rounded rocks and sand, while moraines deposited directly by ice are completely unsorted, a hodgepodge of large and small angular boulders and gravel and sand. Running water always tends to round cobbles and pebbles, and to sort its load by particle size.

Geologists now recognize the Arkansas Valley as the northern end of the Rio Grande Rift, the long thin splinter that subsided when other parts of the Southern Rockies rose during Miocene and Pliocene time. Here the Arkansas River flows along the downfaulted valley, just as the Rio Grande does farther south.

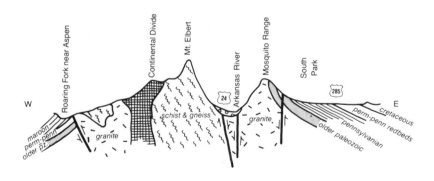

Section across U.S. 24 south of Leadville

Both the Mosquito Range and Sawatch Range are parts of a single Laramide uplift, together forming one wide faulted anticline. Sedimentary rocks on the Mosquito Range dip east; those on the Sawatch Range dip west. So this northern end of the rift is superimposed on the earlier structure, and you are now **in** the central core of the original uplift. Under the stream deposits is Precambrian granite 1750 million years old; it surfaces near the town named — what else? — Granite.

Most of the Sawatch Range peaks are made of Precambrian gneiss and schist. The boundary between these metamorphic rocks and the granite of the Mosquito Range is somewhere beneath the valley fill, probably along hidden faults edging the rift valley wedge.

The Mosquito Range, much lower than the Sawatch Range, was not glaciated here near its south end, and does not have the U-shaped valleys, terminal moraines, or large outwash fans seen in the Sawatch Range. Near Granite, enormous masses of debris washed from the melting glaciers in the Sawatch Range forced the Arkansas River to the east side of the valley, where it carved a channel through the granite. An abandoned pre-Ice-Age channel is buried beneath moraine and outwash gravels.

Gravel terraces in the northern part of the Arkansas Valley are immense, especially north of Mile 187 where the granite across the river disappears. On both sides of the valley, terraces merge upstream with glacial moraines. Each terrace level, once a river floodplain, formed during a relatively stable period when rivers were heavily choked with glacial debris. Later dissected as the river, freed of its ice-fed burden, eroded its channel deeper, they record the on-again off-again history of glaciation.

Leadville lies on these terraces in full view of Mt. Elbert and Mt. Massive. The great peaks dwarf huge slag heaps that spread south of the town, waste from Leadville's mines and smelters. Mines and mine dumps pock the mountain slope east of the picturesque mountain metropolis. The Leadville District yielded more than $500 million in silver, lead, zinc, gold, and other minerals. Its mining history began in the 1860s when placer gold was discovered along the Arkansas River. Later, silver-lead ores were found. They were deposited during the Laramide Orogeny as mineral-rich solutions worked along faults and into Devonian and Mississippian limestone layers. Faults in the mining area bring mineral-enriched rocks to the surface stairstep style all the way up the slope above Leadville.

Museums here tell the town's mining history and display many mineral specimens. Mineral hunting on mine dumps is fun, but **great care should be taken, as old mines are by and large dangerous places**. The Leadville area is full of depressions marking collapsed mine tunnels.

North of Leadville the highway crosses hilly terminal moraines of three glaciers that converged here from the three forks of the Arkansas River. Then it skirts the flat bed of a one-time lake dammed by glacial moraines, and climbs toward Tennessee Pass. Hills east of the highway contain a complicated mosaic of fault blocks in which Precambrian, Paleozoic, and Tertiary rocks are all sliced and crushed and jumbled together. The fault zone parallels the highway across Tennessee Pass. Movement along some of the fault slices has ground the rocks into a fine, powdery, clay-like material which miners call **gouge**; watch for it in roadcuts near the pass.

Tennessee Pass separates drainage of the Arkansas to the south and the Eagle River, a tributary of the Colorado, to the north. The Continental Divide runs along Tennessee Pass, crossing here from the Front Range to the Sawatch Range, and here the Mosquito and Sawatch Ranges converge. Northeast of the pass, forming a high ridge, a thick cliff-forming sill of intrusive igneous rock is sandwiched between Pennsylvanian rock layers.

North of the pass the highway again follows the broken, shattered rocks of the Tennessee Pass fault zone. Scores of dikes further complicate the geology here.

The valley widens suddenly at Eagle Park. Carved by glaciers, dammed by a glacial moraine, the mountain basin contains thick lake, stream, and glacial deposits. A Precambrian fault zone several hundred feet wide may have made the rocks more easily eroded here. On cliffs across Eagle Park, Precambrian rocks are overlain by Cambrian quartzite and the whole sequence of Paleozoic sedimentary rocks shown on the diagram. These rocks dip gently southeastward, continuing the Mosquito Range trend.

Below Eagle Park, the Eagle River enters a rocky canyon formed when glaciers diverted the stream from its normal course. The highway follows a pre-glacial channel farther west.

Descending toward Dowd you pass from Precambrian rocks to Paleozoic ones again; the contact is at the north end of the high bridge at Mile 154. The cliff above the bridge is hard quartzite that was the beach of the Cambrian sea. Devonian and Mississippian limestone lie above it. At Gilman, Precambrian and Cambrian rocks are below the town, the town and its mines are in Devonian and Mississippian layers, and Pennsylvanian rocks show up on steep slopes east of the town, above and below the highway.

The Gilman-Redcliff district is Colorado's main source of zinc, but

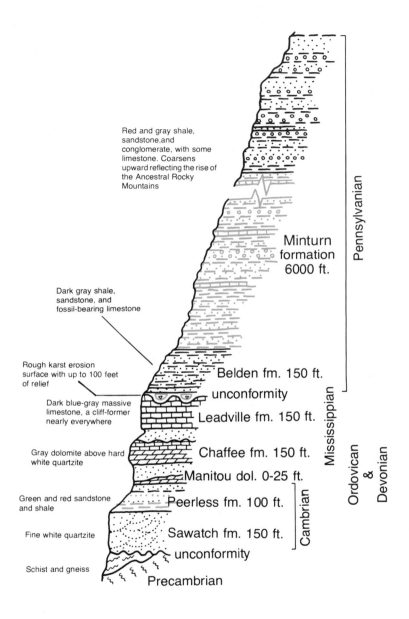

Red and gray shale, sandstone,and conglomerate, with some limestone. Coarsens upward reflecting the rise of the Ancestral Rocky Mountains

Minturn formation 6000 ft.

Dark gray shale, sandstone, and fossil-bearing limestone

Rough karst erosion surface with up to 100 feet of relief

Belden fm. 150 ft.

unconformity

Dark blue-gray massive limestone, a cliff-former nearly everywhere

Leadville fm. 150 ft.

Gray dolomite above hard white quartzite

Chaffee fm. 150 ft.

Manitou dol. 0-25 ft.

Green and red sandstone and shale

Peerless fm. 100 ft.

Fine white quartzite

Sawatch fm. 150 ft.

unconformity

Schist and gneiss

Precambrian

Pennsylvanian

Mississippian

Ordovican & Devonian

Cambrian

Stratigraphic column of rocks exposed along U.S. 24 between Tennessee Pass and Minturn

its riches don't stop there. As at Leadville, most of the ores are in Devonian and Mississippian sedimentary rocks. Zinc occurs in long fingerlike ore bodies near the top of the Mississippian rocks, possibly in one-time caves and caverns in the limestone. Extending from the ends of the fingers down through the Mississippian and Devonian layers are vertical chimneys of copper-silver-lead ores. As if that weren't enough, gold- and silver-bearing veins cut the Cambrian and Precambrian rocks below.

The highway continues through Paleozoic rocks, with the top of the Pennsylvanian in cliffs ahead at Mile 150, the Devonian diving beneath the surface at Minturn.

The 200-foot banded cliffs above Minturn are made up of the Minturn Formation, deposits of debris washed from the Ancestral Rockies and limestone layers formed of the shells of marine organisms. Where these rocks come close to the highway farther north, you can see massive limestone, rich in fossils, deposited as fringing reefs around Frontrangia.

An alternate route between Leadville and I-70, shorter if you're bound for Denver, follows CO. 91 past Climax, the world's largest molybdenum mine. Guided tours of the Climax mill are offered, and exhibits near the parking lot explain the geology of the huge orebody, which lies in a mass of Tertiary igneous rock just east of a major fault zone. The ore is quite low in metal content — only ⅓ of 1% is molybdenum — but the orebody is so large and modern mining methods so efficient that Climax has been a profitable mine. For many years it produced more than half the world's molybdenum.

u.s. 34
loveland — granby
via rocky mountain
national park
(92 miles)

The mountain front west of Loveland shows well the complex folding and faulting that in many places edges the Rockies. Loveland lies on dark gray Cretaceous Pierre Shale that weathers into soil and is rarely exposed. A few miles west of town, older Cretaceous rocks come to the surface where they turn up along the edge of the Front Range.

Near Mile 88 the Dakota Sandstone swoops up into a prominent hogback, with a valley of Jurassic shale and another hogback of older Lyons Formation, pink Triassic sandstone, beyond. Then the Dakota turns down again, vertically, so it appears as a vertical wall called the Devils Backbone. Farther west the sequence repeats once more: Dakota Hogback (Mile 86), Jurassic shale, Triassic Lyons Sandstone. Finally, not as steeply dipping as the Cretaceous rocks, the Pennsylvanian Fountain Formation surfaces by the river. Just beyond it and across another fault, Precambrian rocks wall the canyon of the Big Thompson River.

Before 1976, Big Thompson Canyon sheltered hundreds of pleasant pine-shaded homes and cabins, motels, and picnic areas. The river, fed by high-country streams from Rocky Mountain National Park and the region north of it, rose during the night of July 31, 1976, when heavy rains fell in that area, and swept the canyon with a disastrous flash flood. Homes, motels, cabins, trees, cars and trucks, bridges, a small dam, and U.S. 34 were wiped out by raging waters so savage that they moved twenty-foot boulders and huge blocks of concrete. In spite of flood warnings at least 139 people died, many of them while trying to outrun the flood by driving down the canyon. At the Narrows near the canyon mouth, where it didn't even rain, the stream rose 14 feet in just a few minutes. You'll see evidences of the flood's

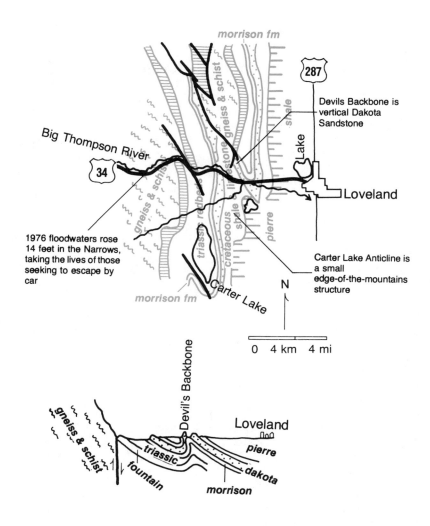

Detail of geology west of Loveland along U.S. 34

rampage as you drive up the canyon, reminders that nature has not finished shaping these mountains. And you'll wonder why so many homes and cabins have been rebuilt close to the river.

The canyon walls are of metamorphic rock, banded gneiss and shiny schist more than 1750 million years old, broken, fractured, and penetrated by pegmatite dikes, some with large mica crystals. Near

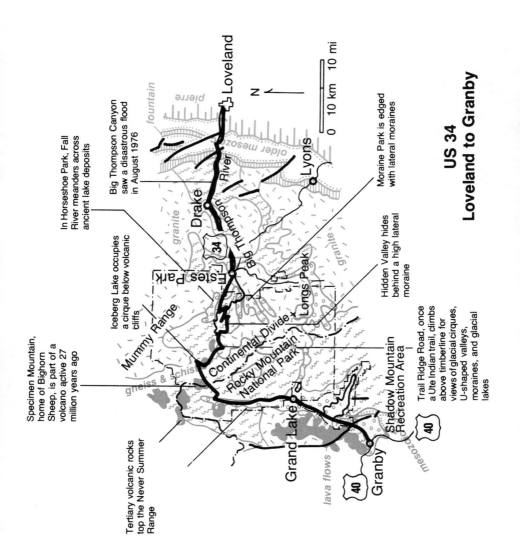

Specimen Mountain, home of Bighorn Sheep, is part of a volcano active 27 million years ago

In Horseshoe Park, Fall River meanders across ancient lake deposits

Big Thompson Canyon saw a disastrous flood in August 1976

Iceberg Lake occupies a cirque below volcanic cliffs

Moraine Park is edged with lateral moraines

Hidden Valley hides behind a high lateral moraine

Tertiary volcanic rocks top the Never Summer Range

Trail Ridge Road, once a Ute Indian trail, climbs above timberline for views of glacial cirques, U-shaped valleys, moraines, and glacial lakes

Loveland

Drake

Estes Park

Lyons

Mummy Range

Longs Peak

Continental Divide

Rocky Mountain National Park

Grand Lake

Shadow Mountain Recreation Area

Granby

Big Thompson River

pierre

fountain

older mesozoic

granite

granite

gneiss & schist

lava flows

mesozoic

N

0   10 km   10 mi

**US 34
Loveland to Granby**

34

40

40

34

the upper end of the canyon these rocks alternate with patches of pink and dark red granite until finally granite predominates. It weathers quite differently than the metamorphic rocks, forming rounded boulders and domes rather than craggy cliffs. The canyon widens, and the highway emerges onto rolling granite hills that merge into Estes Park in the heart of the Front Range.

The Twin Owls tower over the peaceful valley of Estes Park, sparsely settled when this photograph was taken in 1916. These and other granite domes form because granite expands slightly when freed of overlying rock, causing a set of joints paralleling large exposed surfaces. Weathering processes then work along the joints, and freezing and thawing spall off curving slabs of rock.

W.T. LEE PHOTO, COURTESY OF USGS

Though Estes Park with its border of high granite domes looks like a wide glacier-carved valley, it is below the lower limit of Ice-Age glaciation and is really a deep stream-eroded canyon like the one you have driven through, filled in with thick deposits of gravel and rock. Lake Estes, man-made, serves as a staging point for western-slope water brought by tunnel under Rocky Mountain Park and the Continental Divide to the eastern slope. The tunnel portal can be seen across the valley.

There are two east entrances to Rocky Mountain National Park. Stay on U.S. 34 for this itinerary, or take Colorado 66 to the Visitor Center and Moraine Park. The routes rejoin.

Above Estes Park, U.S. 34 is once more confined within a narrow canyon carved in Precambrian granite. In a few miles the canyon

As seen from the Fall River road, sediments deposited in moraine-dammed lakes flatten the floor of the U-shaped glacier-cut valley of Fall River.

JACK RATHBONE PHOTO

opens into a glacial valley almost precisely at the 8000-foot elevation considered the lower limit of Pleistocene glaciation in Colorado. Three glacial stages are represented: You enter the valley across the lowest, oldest terminal moraine, so deeply weathered and decayed that it is hardly recognizable. Just below the checking station is a moraine from the second glacial episode, covered with gray-brown soil and dark with pines and spruce. The moraine of the last glacial stage is half a mile beyond the checking station, where the highway curves and begins to climb. Note the sandy soil on this moraine and the fresh-looking boulders, probably deposited less than 10,000 years ago.

Horseshoe Park above this moraine became a lake as the last glacier melted; it is floored with lake-deposited mud and sand.

At the head of Horseshoe Park you can take the old Fall River Road (unpaved), following the nature guide available where it begins to climb. Many geologic features are described. Or continue on U.S. 34 with this itinerary. Routes rejoin at Fall River Pass.

149

Beyond the junction with Colorado 66, U.S. 34 becomes Trail Ridge Road, following an old Ute trail across the mountains. Views of Horseshoe Park and of Moraine Park to the south show that the trail climbs a narrow ridge between two broad glaciated valleys. Picture these green parks as they were 10,000 years ago, filled with creeping tongues of ice cut by arcuate crevasses and streaked with rock and sand, terminating at high-piled rocky moraines. Frigid winds blew downslope from icy peaks to the west, and little vegetation — only low-growing tundra plants and willows — grew near here.

Looking down on Fall River in Horseshoe Park you can see that it meanders now in tight loops on the flat valley floor, for there isn't much gradient to tell it which way is downhill. For more than a mile Horseshoe Park is hidden from view by a long lateral moraine (the one that also hides Hidden Valley).

The core of the Front Range exposed along Trail Ridge Road is mostly gneiss and schist, metamorphic rock like that in Big Thompson Canyon. In places it is intruded by younger granite. Granite forms Longs Peak towering to the south. Some geologists think that this peak's flat top is a remnant of the flat erosion surface formed at the end of Precambrian time, 650 million years ago. (In Glenwood Canyon such a surface is covered directly by Cambrian rocks.) Or the

Ice Age glaciers scooped out the cirque and U-shaped valley of Loch Vale in Rocky Mountain National Park.
JACK RATHBONE PHOTO

150

*Above timberline along Trail Ridge Road, even today's climate is
arctic. Trees cannot grow, and short-seasoned midsummer
wildflowers toss in chilly winds. Freezing nights and frigid winters
leave other marks: shattered rocks broken and tipped at drunken
angles, barren jagged surfaces known as felfields, and patterned
ground marked by constant frost-heaving of soil and stones.*

W.T. LEE PHOTO, COURTESY OF USGS

surface may have formed as the Ancestral Rocky Mountains were
eroded away 275 million years ago.

As the highway climbs higher, many other glacial features become
visible: a broad glaciated upland cut by glaciated cliffs; cirques,
cirque lakes, and hanging valleys; even some small "living" glaciers,
distinguished from snowfields by the lip-like piles of their terminal
moraines.

From Forest Canyon Overlook, a five-minute walk, look down 2500
feet into Forest Canyon, a steep-walled glacially gouged trough.
Before the Ice Ages, the Big Thompson River probably followed as
tortuous a course here as in its lower reaches today. But glaciers have
trouble negotiating turns, so they straighten the valleys in which
they flow, grinding off spurs and smoothing curves.

Across the canyon little lakes stairstep up hanging valleys, filling
hollows scooped from bedrock by now-vanished glaciers. At the head
of Hayden Creek, named for a pioneer American geologist, an icefield
nestles in a small cirque once occupied by a glacier.

Stop at Fall River Pass to look down into the cloverleaf cluster of
three cirques that head Fall River Canyon. Exhibits in the Visitor
Center explain many alpine geologic and biologic features.

Northwest across the Cache La Poudre Valley from Medicine Bow Curve rises red-tinted Specimen Mountain, the only remnant of a volcano in the park. It was active around 27 million years ago, and at that time was certainly much higher and probably much more conical. Soft yellowish ash from its eruptions is visible in roadcuts near Poudre Lake.

In this lake, biologic processes work today to accomplish geologic ends. Dammed by a moraine and a small beaver dam, the lake is filling with marsh grass and other vegetation and will eventually become a mountain meadow. The peat-like material that is filling it may someday become coal.

At Farview Curve, look down into the Kawuneeche Valley where the young Colorado River meanders lazily through a long, straight, glaciated valley occupied in Ice-Age time by a succession of glaciers. From glacial meltwater the stream gathered strength and sped southwest, joining other rivers and streams, to carve Glenwood Canyon, Glen Canyon, and ultimately Grand Canyon in Arizona. Across the valley decomposing igneous rocks in the Never Summer Range add a touch of brilliance to the landscape. Iron minerals cause the color, though there is not enough iron for profitable mining.

Grand Lake, half a mile east of the road, is a natural lake dammed by the lateral moraine of the former Kawuneeche Valley glacier and the terminal moraine of a glacier that came down Paradise Creek. Shadow Mountain Reservoir and Lake Granby, man-made reservoirs, store Colorado River water that is eventually tunneled to Lake Estes and east slope cities. Mountains west of the river here are topped with Tertiary volcanic rocks; they and broad, high river terraces bury most of the band of upturned sediments we would otherwise expect to find on this side of the Front Range.

Semicircular cirques carved by Pleistocene glaciers frame Berthoud Pass in this aerial photograph. James Peak is at the upper right; Fraser River Valley is beyond the pass.

u.s. 40
empire — kremmling
(73 miles)

Between Empire and Winter Park, U.S. 40 zigzags up over Berthoud Pass and down again into the valley of the Fraser River, crossing the Continental Divide and a lot of Precambrian rocks in the process. This is avalanche country, for steep slopes and deep snow are a precarious mix. When packed snow breaks loose and whooshes down the mountainside, it gathers up more snow and grows larger and more dangerous. Broad swaths overgrown with stands of young trees mark old avalanche paths. The Colorado Highway Department now watches and measures snowpack and dynamites potential avalanches.

This is also glacially eroded country, and several moraines cross the U-shaped valley of the West Fork of Clear Creek. As the highway switchbacks up toward Berthoud Pass, look across the West Fork at two high bite-shaped cirques on Woods Mountain. Little glaciers that scooped out these cirques were tributaries to the larger glacier in the main valley. Rough rocky moraine piles below each cirque show that small glaciers survived there for some time after the main glacier began to shrink.

153

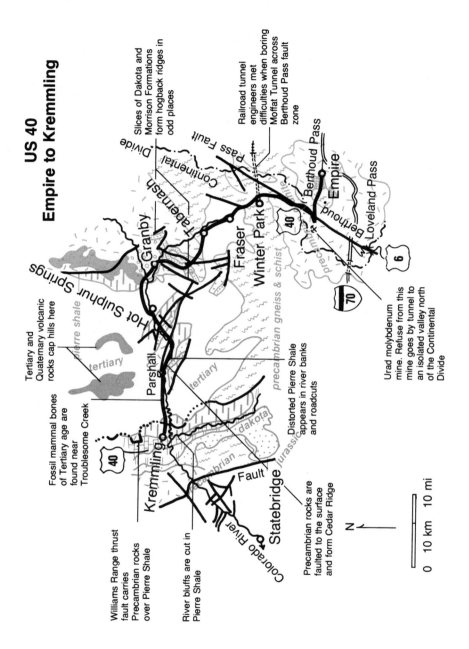

US 40
Empire to Kremmling

Slices of Dakota and Morrison Formations form hogback ridges in odd places

Railroad tunnel engineers met difficulties when boring Moffat Tunnel across Berthoud Pass fault zone

Continental Divide

Berthoud Pass Fault

Tabernash

Granby

Berthoud Pass

Empire

Loveland Pass

Berthoud

Fraser

Winter Park

40

precambrian gneiss & schist

precambrian

6

70

Urad molybdenum mine. Refuse from this mine goes by tunnel to an isolated valley north of the Continental Divide

Tertiary and Quaternary volcanic rocks cap hills here

Hot Sulphur Springs

Pierre shale

tertiary

Parshall

tertiary

dakota

Jurassic

Distorted Pierre Shale appears in river banks and roadcuts

Fossil mammal bones of Tertiary age are found near Troublesome Creek

40

Kremmling

cambrian

Statebridge

Fault

Williams Range thrust fault carries Precambrian rocks over Pierre Shale

River bluffs are cut in Pierre Shale

Colorado River

Precambrian rocks are faulted to the surface and form Cedar Ridge

N

0   10 km   10 mi

0   10 mi

Berthoud Pass, a notch in the crest of the Front Range, marks the point where a fault zone crosses the divide, weakening the hard Precambrian rocks so they erode more easily. The fault also crosses Loveland Pass 11 miles southwest of here. Seven miles northeast, the same fault caused collapse problems in boring Moffat Tunnel, where the main line of the railroad goes through the Continental Divide.

All the rocks seen from the pass are Precambrian. There is gray granite on both sides of the pass, but a narrow finger of metamorphic rock outcrops at the pass itself. In roadcuts you can see the intense chaotic fracturing near the fault; the rock is completely shattered and slivered. On the other side of the pass, the Fraser River heads in this fractured zone.

The Fraser River Valley is floored by glacial moraine — also in its way chaotic, with humps and bumps enclosing small ponds and marshy depressions. Near Winter Park, where well developed terminal moraines cross the valley, notice the unsorted nature of the moraine material — sand, clay, and boulders of all sizes jumbled together.

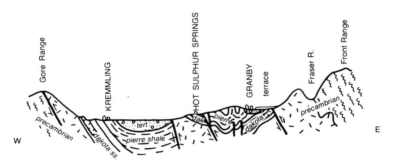

Section from Front Range to Gore Range through Granby

Below Winter Park, the valley is floored with Tertiary rocks concealed by **glacial outwash**, stream-deposited glacial debris. Some of the Tertiary sediments are exposed in ridges across the valley, muddy brown sandstone that contains a good deal of volcanic ash, some pure volcanic ash layers, and several mud and gravel layers. High flat-topped terraces in the valley near Fraser and Tabernash are made of glacial outwash deposited by streams below melting glacial tongues. Beyond Tabernash there is no further evidence of glaciation from the Fraser River direction.

Swinging west the road goes through hills of Tertiary sediment interspersed with purple granite. Careful geologic mapping has shown that the Precambrian granite is thrust westward over Cretaceous rocks, and that they in turn are thrust over Tertiary strata, apparently by expansion of the uplifted granite. In this area so many faults go in so many directions that little blocks of this and that rock pop to the surface where you least expect them. The Dakota Sandstone is the most recognizable unit, often adjacent to colorful green and purple Morrison Shale. Where the structures are complex, about the best you can do is try to spot Precambrian granite and Cretaceous Dakota Sandstone, the latter usually in hogbacks like the one east of the Front Range. Not all the complications are shown on the map.

The high terrace behind Granby is outwash from a large glacier that crept south down the long valley of the Colorado River, receiving input from smaller glaciers in Rocky Mountain National Park. West of Granby the road enters a narrow canyon bordered by cliffs of dark volcanic rocks.

At Hot Sulphur Springs, hot springs rise where water heated well below the surface comes up rapidly along a fault. The springs are highly mineralized, and of all the minerals the smelly sulphur is the most noticeable. Dakota hogbacks east and west of town are little fault blocks. Going west, you'll see Jurassic Morrison Shale lying between the Dakota and Precambrian rocks. This region was part of the Ancestral Rockies, and Paleozoic rocks were lifted and washed away before the Jurassic shale was deposited.

Just west of Hot Sulphur Springs, the Colorado River enters a canyon between granite on the north faulted over Tertiary sediments on the south. The fault is at the base of the coarse red granite north of the road. Across the river, another block of granite forms a ridge with the Morrison and Dakota Formations leaning up against it. Emerging from the canyon, the river and highway are at last out of the complexly faulted area and on the relatively featureless surface of the much younger Tertiary sediments that floor Middle Park.

Because of all the faulting along its fringes, Middle Park is not as coherent a geographic unit as it might be. Its margins are ragged, and various hills and ridges poke above the Pleistocene terraces or Tertiary sediments of its floor, as you have seen. The faults that border it are extensions of the same ones that define North Park, but a belt of volcanic rocks — the Rabbit Ears Range — separates them topographically.

The Tertiary sediments are composed of brown sand and gravel and stream-rounded boulders washed westward off the Front Range,

In Middle Park, the Pierre Shale includes a sandstone unit which caps the bluffs near Kremmling. Cretaceous marine fossils, particularly the straight-shelled ammonite Baculites, have been found in the shale slopes below these bluffs.

G.A. IZETT PHOTO, COURTESY OF USGS

eastward off the Park Range, and northward off the Gore Range. Notice how poorly consolidated these rocks are — you can hardly call them rocks. Sand grains can be scraped off with a fingernail. In general, sedimentary rocks harden with age, so these young Tertiary rocks are only lightly glued together; erosion cuts into them quite easily. The Quaternary gravels of the outwash terraces are even more easily washed away by the river.

Just east of Kremmling muddy brown Cretaceous Pierre Shale comes to the surface, the same shale that underlies Denver and much of the Piedmont east of the Front Range. This shale is well exposed in bluffs behind Kremmling. The Colorado River turns southwestward abruptly at Kremmling and cuts through the hard Precambrian core of the Gore Range-Park Range uplift. You can see its deeply incised canyon to the south. Both the Gore Range and the Park Range are faulted anticlines, like most Colorado mountains, but the Gore Range is unusual in that fragments of some of the sedimentary rocks that arched across its core are still in place on top of the range.

157

The line of a thrust fault shows up clearly on Wolford Mountain, where trees grow on coarse Precambrian granite in the top half of the mountain but don't take root in the Cretaceous shales below. Thirty or forty miles south the same fault delineates the west margin of the Front Range.   JACK RATHBONE PHOTO

u.s. 40

kremmling — steamboat springs

(52 miles)

Leaving Kremmling, U.S. 40 runs north along a shallow valley underlain by Cretaceous limestone and shale. The highway is more or less parallel to a long thrust fault that brings ancient Precambrian rocks west over much younger shale. The line of the thrust fault is usually concealed by soil and Tertiary sediments east of the highway.

Near and north of Wolford Mountain, the Cretaceous shale commonly contains marine fossils: ammonites that are ancient relatives of the Chambered Nautilus, high-spired snails, and big corrugated clams called *Inoceramus*.

Whiteley Peak, a prominent pointed peak about 15 miles north of Wolford Mountain, is capped with basalt that shows the columnar jointing often characterizing volcanic rocks. Vertical shrinkage cracks that open as lava cools and shrinks form polygonal columns.

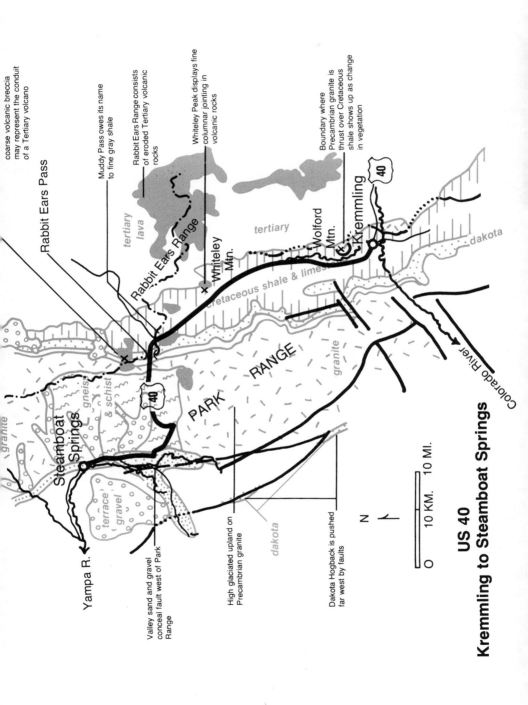

coarse volcanic breccia may represent the conduit of a Tertiary volcano

Rabbit Ears Pass

Muddy Pass owes its name to fine gray shale

Rabbit Ears Range consists of eroded Tertiary volcanic rocks

Whiteley Peak displays fine columnar jointing in volcanic rocks

Boundary where Precambrian granite is thrust over Cretaceous shale shows up as change in vegetation

tertiary lava

Rabbit Ears Range

Whiteley Mtn.

tertiary

Wolford Mtn.

Kremmling

40

cretaceous shale & limestone

dakota

PARK RANGE

granite

gneiss & schist

Steamboat Springs

40

granite

terrace gravel

dakota

Colorado River

Yampa R.

Valley sand and gravel conceal fault west of Park Range

High glaciated upland on Precambrian granite

Dakota Hogback is pushed far west by faults

N

0   10 KM.   10 MI.

US 40
Kremmling to Steamboat Springs

At Muddy Pass U.S. 40 swings west and begins the climb to Rabbit Ears Pass at the top of the Park Range. Beaver dams just beyond Muddy Pass are reminders that beavers are active geologic agents, damming streams, reducing floods, and eventually creating new mountain meadows. In this area the bumpy irregular terrain is the surface of a large glacial moraine, probably made hummocky by landsliding. Several even larger moraines on the east side of the Park Range are so hummocky that they harbor small hidden ponds and lakes without surface drainage.

Do you recognize the Dakota Sandstone as the highway climbs toward the pass? It extends to the summit of the Park Range, where it lines up with the Rabbit Ears themselves. From the highway you get good views of the Park and Gore Ranges, North and Middle Parks, and the Front Range in the distance to the east. On clear days you can see light-colored sand dune fields in North Park.

Rabbit Ears Peak, with its cluster of three summits, is a volcanic plug, the lava-filled feeder pipe of a former volcano. It is made of a rough, coarse volcanic rock called **agglomerate**, containing chunks of broken volcanic material cemented together with reddish lava.

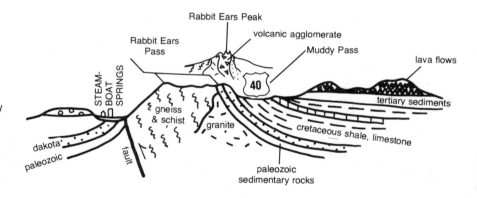

Section across the Park Range at Rabbit Ears Pass

West of Mile 152 on Rabbit Ears Pass is a thin band of dark red granite, and then gneiss and schist that make up most of the core of this part of the Park Range. These metamorphic rocks are decoratively swirled like a marble cake with light gray and dark gray bands, all laced by a network of pegmatite dikes. Chances are that these swirls came into existence through a sort of semi-fluid convective movement that stirred together half-melted porridge-like rock material of different composition. Such swirling metamorphic rocks form at temperatures and pressures high enough to bring them to the verge of melting. Originally they were probably sedimentary rocks, sandstone and shale; although they look much different now they still retain the chemical composition of these sedimentary rocks.

The Park Range was once occupied by glaciers. Its flat uplands supported several icecaps in Pleistocene time, with fingers of ice reaching down the slopes and into the canyons to deposit large moraines. North of Rabbit Ears in the Mt. Zirkel Wilderness Area small perennial snowfields exist now; they probably do not move enough to be classed as glaciers. Most geologists and climatologists feel that they are young snowfields that started from scratch about 4000 years ago, after a warmer climatic cycle that completely melted the Ice-Age glaciers. Valleys on both sides of the range head in steep-walled cirques and terminate in extensive moraines. Parts of the flat-floored Yampa Valley near Steamboat Springs are spangled with big boulders washed from these moraines during flash floods.

The west side of the Park Range is steep and dramatic. Faults that edge it north and south of Steamboat Springs almost certainly pass beneath the sand and gravel of the Yampa River Valley near the highway. There appears to be quite an offset in the Dakota Hogback here.

Hot springs at Steamboat Springs occur along both banks of the Yampa River. Some are mere seeps but others are active bubbling pools. One of them used to make a noise like the chugging of a steamboat, giving the town its name, but the sound ceased when a railroad cut was excavated near by. The hot springs issue from the Dakota Sandstone where water heated by its passage through hot rocks below the surface rises rapidly along faults. Along the banks of the Yampa, and on the slopes of Quarry Mountain to the southwest, are deposits of dull gray travertine. Highly mineralized hot water, charged with calcium carbonate dissolved from limestone, evaporates at the surface and leaves the calcium carbonate behind in this form.

u.s. 50

canon city — poncha springs

(62 miles)

Leaving Canon City, U.S. 50 loops around the southern end of the Dakota Hogback, then runs north beside it. The Dakota and other Mesozoic formations dip steeply here. For a better look at them take 2½-mile Skyline Drive along the crest of the hogback. From there you can see the line of the hogback coming up from the south, passing Canon City, and curving north around Garden Park. Resistant Cretaceous limestones form parallel but less pronounced ridges. Some of the sandstone layers in the Dakota Formation along the top of the ridge are patterned with ripple marks, mudcracks, and tracks and trails of ancient lagoon-dwelling animals, suggesting that they were deposited on beaches and sand bars bordering the one-time sea.

To the west, quarries visible from Skyline Drive are in Ordovician rock, the Harding Sandstone and Fremont Dolomite. In one of these quarries, in 1887, paleontologist Charles Walcott discovered bony plates of extinct armored fish called *Astraspis,* previously known only from much younger Devonian rocks. Walcott identified enough associated corals and shellfish to be sure of the age of the rocks, and realized he had found the world's oldest known vertebrate animals, deposited here 440-500 million years ago. Not until 1977 were yet older fossil fish discovered, in Antarctica and Wyoming.

The highway crosses the Harding Sandstone as it curves westward a few miles from Canon City. A thin Ordovician limestone lies right on coarse pink Precambrian granite, as in roadcuts between Miles 273 and 272.

Royal Gorge, where the Arkansas River cuts into Precambrian rocks (5-mile side trip) is 1200 feet deep, one of the deepest canyons in Colorado. The road to Royal Gorge Bridge crosses Precambrian granite and then enters a band of brownish gray metamorphic rock that resembles the Idaho Springs Formation near Idaho Springs, west of Denver. Here, this rock is intruded by many large pegmatite dikes containing big crystals of quartz, feldspar, and white mica or muscovite. "Books" of mica four or five inches across have been quarried

162

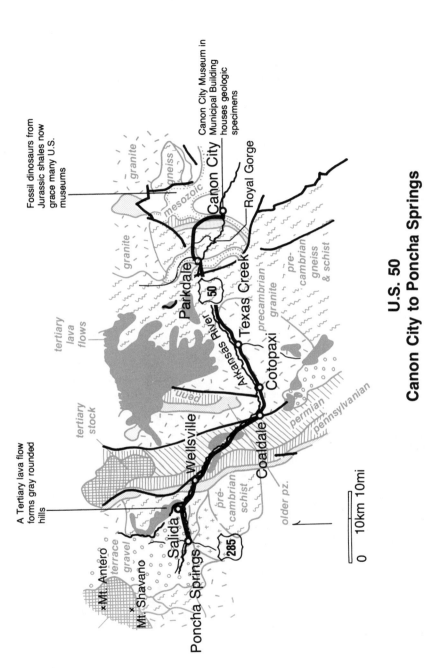

Fossil dinosaurs from Jurassic shales now grace many U.S. museums

Canon City Museum in Municipal Building houses geologic specimens

A Tertiary lava flow forms gray rounded hills

U.S. 50
Canon City to Poncha Springs

163

here; big chunks can be seen in pegmatite used as lane markers near Royal Gorge Bridge. Mica, a poor electrical conductor, is used as an insulator in electrical equipment; it also lends its sparkle to wallpaper, Christmas tree snow, roofing paper, and paint. Crystals of microcline feldspar, masses of snow white quartz, and large crystals of beryl, black tourmaline, and garnet occur here too, though not of gem quality. Nearby granite sometimes contains intergrown quartz and feldspar crystals that look like some ancient form of writing; rock like this is called **graphic granite**.

Right at Royal Gorge there is a curious mixture of gneiss, schist, and granite in which the granite appears to have been injected along

Royal Gorge was cut by the Arkansas River through a relatively small wedge of Precambrian granite and metamorphic rocks that extends south from the Pikes Peak massif. The river course, established on Cretaceous and Tertiary rocks before Miocene-Pliocene uplift, cuts directly across the once buried, now denuded hills of Precambrian rock.

M.R. CAMPBELL PHOTO,
COURTESY OF USGS

164

the flaky cleavage planes of the gneiss and schist. Some granite masses are hundreds of feet across; others are half an inch thick. Geologists refer to this type of rock as **injection gneiss** or **migmatite.**

The even upper surface of these ancient rocks, as seen along the rim of Royal Gorge, is part of an ancient peneplain, perhaps formed during the long-drawn-out erosional interval at the end of Precambrian time, or perhaps due to several stages of erosion between Precambrian and Jurassic time, when the overlying Morrison Formation was deposited.

West of Royal Gorge and back on U.S. 50, near Parkdale, there is a small patch of badly faulted and shoved-around rocks. Precambrian rocks were faulted repeatedly here, and wedges of red and yellow Mesozoic rocks, most of them dipping west but some vertical and some even upside down, border the river and the highway. Paleozoic rocks are lacking, for this region was part of the Ancestral Rocky Mountain highland called Frontrangia, and older Paleozoic rocks were eroded away.

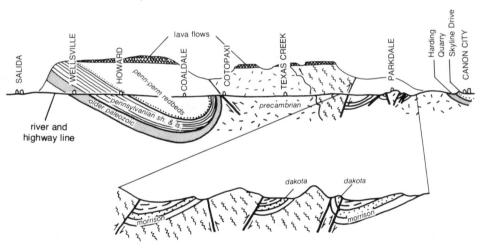

Section along U.S. 50, Canon City to Salida

West of Parkdale, Precambrian rocks cut by black and white dikes again appear. Then once more at Mile 257 there are injections of pink granite, with a typical way of weathering into bare, rounded slopes. These rocks are well exposed where the canyon deepens near Cotopaxi. The dominant joints or fractures parallel the trend of a Precambrian mountain range that existed here 1.7 billion years ago. North across the river at Mile 257 the granite hills are capped with

165

lava flows from the Thirtynine-Mile Volcanic Field at the south end of South Park. Metamorphic rocks, especially black hornblende gneiss, recur from place to place.

Just east of Coaldale a striking change in scenery is brought about by the Pleasant Valley fault, which extends far northward and separates Precambrian rocks from dark red sandstone and shale of Permian and Pennsylvanian age. These steeply dipping rocks formed from the sand and mud, pebbles and cobbles washed westward off the west side of Frontrangia. Near Coaldale they contain coal (as you might guess) and gypsum, which is quarried west of the town. Fine grains of iron oxide — the mineral hematite — give them their red color.

Where they first appear (Mile 238) the red sediments are dragged upward and overturned by the upward movement of granite east of the big fault. For a few miles the fault runs right up the bed of the Arkansas, or to put it more accurately the Arkansas runs right along the line of the fault, and rocks to the northeast are finely fractured granite.

About 20,000 feet of Pennsylvanian and Permian redbeds occur here, as measured perpendicular to the beds or layers. They may be doubled up by faulting, but if so the faults are parallel to the bedding and are virtually impossible to detect. Few fossils have been found in the sandstone or shale, though fossils are present in some limestone beds. They tell us these rocks are Permian at the top (east) and grade gradually downward (westward) into Pennsylvanian strata.

Upriver near Mile 232 the shales become more abundant and thicker and the sandstones are supplanted by gray limestone; here the rocks are definitely Pennsylvanian. Enough coal beds are interspersed to show that the Pennsylvanian rocks were deposited in nearshore lagoons, bays, and swamps. In places the rocks are clearly cyclic; that is, there are often-repeated sequences of sandstone-shale-sandstone-shale (as at Mile 231) or limestone-shale-limestone-shale. The environment in which they were deposited must have fluctuated in some way, regularly and repeatedly. All over the world cycles characterize Pennsylvanian rocks. Numerous geologists have tried to explain the worldwide cycle patterns, and the most likely explanation seems to be the idea that sea level varied a great deal, perhaps because of variations in the pace of sea floor spreading or perhaps in response to often-repeated glacial episodes somewhere where they have left little or no record.

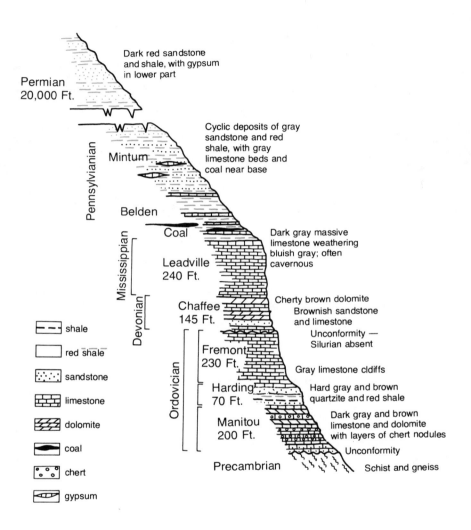

Stratigraphic column of rocklayers exposed near Wellsville, in Arkansas Canyon. Symbols indicate rock type.

*Northwest of Salida, Mt. Princeton and other Sawatch Range peaks
top 14,000 feet. Miocene-Pliocene sediments sloping from the
mountains record the regional uplift; they are capped by coarse
Pleistocene gravels that reflect both uplift and glaciation of the
mountain area. The light zone on Mt. Princeton's flank is an area
leached by hot springs.* R.E. VAN ALSTINE PHOTO, COURTESY OF USGS

West of the Pennsylvanian rocks, just east of Mile 228, is a massive
gray cliff of Mississippian Leadville Limestone, very widespread and
probably once continuous with similar gray limestone found from
Arizona to Alberta. Farther west are brownish Devonian limestone
(near the Chaffee County line) and Ordovician limestone, sandstone,
and shale (across the river at Mile 227). Quarries near Mile 227
obtain travertine from recent or Pleistocene hot spring deposits for
use as soil conditioner.

Some of the hills near Salida, particularly soft, rubble-covered hills
like the one with the S, are Tertiary intrusions and lava flows. Others
are Precambrian granite. West of town, large pediments and alluvial
fans slope from the Sawatch Range on the other side of the valley.
This range contains many of Colorado's highest peaks. The three at
its southern end, Mt. Shavano, Mt. Antero, and Mt. Princeton (all
Fourteeners), are eroded from a single Tertiary batholith. The
Sawatch uplift includes the Arkansas Hills east of the river as well as
the much higher Sawatch Range. The Arkansas River flows along a
slim downfaulted block in the core of the uplift.

u.s. 50
poncha springs — gunnison

(62 miles)

West of Poncha Springs, U.S. 50 follows the South Fork of the Arkansas River, climbing gradually across deposits of Pleistocene gravel. Shavano Peak and Mt. Antero, both Fourteeners, rise in the Sawatch Range to the north. These peaks, as well as Mt. Princeton farther north, are parts of a small Tertiary batholith. More or less circular in outline, the batholith is 15 to 20 miles across.

West of Maysville, the highway abruptly enters Precambrian rocks (Mile 210), light and dark gray gneiss with some schist, the latter recognizable by its dark color and platy way of breaking. Gray granite is exposed here too, as in the massive cliff above Mile 206. Note the abundance of coarse gravel near the road, with rounded boulders indicating that it was stream-deposited, probably by streams originating in glaciers during the Ice Ages.

Farther west the highway passes through a steeply tilted wedge of Paleozoic sedimentary layers similar to those east of Salida in the Arkansas River Canyon, described in the preceding section. Watch first for exposures of dark red Ordovician limestone. They lie right on the granite but the contact can't be clearly seen. The erosion surface on the granite below the Ordovician limestone represents the immensely long Lipalian Interval plus the 100-million-year span of Cambrian time. Cambrian rocks were apparently never deposited here, perhaps because this region was an island in the Cambrian sea.

Cliffs above the highway at Garfield are Paleozoic limestone. A smooth, shiny rock surface, polished by movement of rock against rock, can be seen on this limestone just across the river from Garfield. Steeply tipping rocks in the huge quarry near Monarch are Leadville Limestone, a gray Mississippian rock that probably corresponds to similar massive gray limestone in adjacent states. Broad surfaces

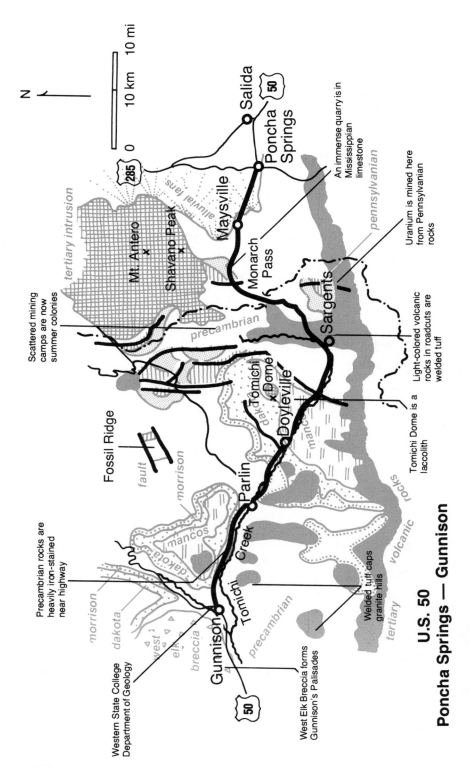

N

0   10 km
0   10 mi

285

Scattered mining camps are now summer colonies

tertiary intrusion

Mt. Antero x

Shavano Peak x

precambrian

alluvial fans

Maysville

Salida

50

Poncha Springs

Monarch Pass

An immense quarry is in Mississippian limestone

pennsylvanian

Uranium is mined here from Pennsylvanian rocks

Sargents

Light-colored volcanic rocks in roadcuts are welded tuff

Precambrian rocks are heavily iron-stained near highway

Fossil Ridge

fault

morrison

mancos

dakota

dakota

Tomichi Dome x

Doyleville

Parlin

Tomichi Creek

mancos

Tomichi Dome is a laccolith

morrison

mancos

dakota

west 1

elk breccia

Gunnison

precambrian

volcanic rocks

tertiary

Welded tuff caps granite hills

Western State College Department of Geology

50

West Elk Breccia forms Gunnison's Palisades

U.S. 50

Poncha Springs — Gunnison

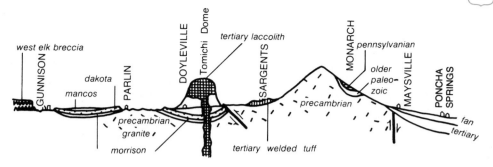

Section along U.S. 50 from Poncha Springs to Gunnison

sloping steeply toward you in the quarry are surfaces of individual limestone beds which make convenient quarrying planes. The limestone goes to the Colorado Fuel and Iron Company smelters at Pueblo, where it is used as a furnace flux: mixed with iron ore it helps to convert earthy impurities in the red hot, molten ore into liquid **slag**, a foamy, lava-like material which can be floated off and discarded. Almost opposite the quarry entrance, Pennsylvanian shale appears in a roadcut to the right. Above the quarry is the Madonna Mine, which used to produce lead, zinc, and silver. Mine tours are conducted here; they go about 1500 feet into the mountainside.

Between the quarry and Monarch Pass and for some distance down the west side of the pass, light gray Precambrian granite appears again. Notice the rusty iron oxide on fracture faces. This area is within the Colorado Mineral Belt, the 50-mile-wide mineralized zone extending from Boulder County to southwestern Colorado. Within this belt, such iron-staining is common.

Looking north from Monarch Pass, Mt. Aetna (the most conspicuous nearby peak) is part of the same Tertiary batholith as the peaks farther north. High granite country south of Monarch Pass is capped in the distance with Tertiary volcanic rocks, mostly of a kind called welded tuff, that spread northeastward from eruptions in the San Juan volcanic area. Mt. Ouray, a truncated pyramid that rises above the volcanic rocks, marks the south end of the Precambrian core of the Sawatch Uplift.

The valley on the west side of Monarch Pass was never glaciated. Note its V-shaped canyon profile, and the deeply weathered rock that would have been the first thing scraped off by glaciers. At the foot of the pass, however, the highway crosses the moraine of a glacier that flowed down Tomichi Creek valley nearly to Sargents.

Along both sides of this part of the Sawatch Range there are little splinters and wedges of Paleozoic rock caught between faults in the granite. They usually contain Ordovician, Devonian, Mississippian, and Pennsylvanian sedimentary rocks. You saw one of them at Monarch quarry — the only one the highway passes through. Many of the splinters are mineral-enriched and were centers of mining activity in the 1870s, 80s, and 90s. Ghost towns abound in this region — most are now summer communities — and Sargents was once a bustling railroad town supplying them. East of Sargents a more recently developed mine, not visible from the highway, obtains uranium from Pennsylvanian strata in another Paleozoic wedge.

West of Sargents along Tomichi Creek, Precambrian rock includes both granite and dark hornblende gneiss, cut by many dikes. Several dikes are well exposed in roadcuts; dark ones are similar in composition to dark gray basalt; light ones are similar to granite. Also visible in these cuts are **xenoliths** of wall rock (in this case, gneiss) broken off and "floating" in the granite. Intruded as a batholith 1450 million years ago, the granite often fingered its way into a fairly close relationship with the gneiss.

Just west of Mile 182, Precambrian rocks of the Sawatch Range core override the Cretaceous Dakota Sandstone, which has been bent up, dragged, and turned upside-down by movement on the fault there. The Dakota Sandstone makes a vertical cliff visible about a mile north of the highway. Between this fault and Parlin the highway crosses a saucer-shaped basin of Mesozoic rocks, with a hogback of Dakota Sandstone rimming a broad sage-covered valley of the much softer Mancos Shale.

One of the most striking landmarks in this area is round-topped Tomichi Dome. It is a laccolith formed when gummy, viscous molten rock was forcibly injected between sedimentary layers, spreading sideways between some layers and doming up the layers above. It went up through the Dakota Sandstone and Mancos Shale but bowed up younger strata, most of which are now eroded away.

West of Tomichi Dome the route enters an area which in late Paleozoic and early Mesozoic time was part of Uncompahgria, the western of the two island ranges of the Ancestral Rockies. The crest of the old highland was somewhere near the present town of Gunnison. Because this area was being eroded during that time, no Paleozoic or Triassic sedimentary rocks remain here; Jurassic strata lie directly on the Precambrian surface. (A few miles north, along Fossil Ridge, Paleozoic rocks are well developed, with fossils galore.)

Outcrops of pink granite near Parlin weather along three intersecting sets of joints, and separate into boulders. Rounding occurs by weathering of protruding corners, and by a process known as spalling, in which large curving rock flakes peel off the boulders as in the center of the picture. M.H. STAATZ PHOTO, COURTESY OF USGS

Among Mesozoic formations three are quite distinctive and easily recognized. From top to bottom, they are:

• Mancos Shale, a slope-forming black Cretaceous shale that weathers light gray or yellow.

• Dakota Sandstone, a resistant Cretaceous sandstone that forms cliffs where it is horizontal, hogbacks where it is tilted, and high walls where it is vertical.

• Morrison Formation, colorful greenish and reddish Jurassic shale, another slope-former.

The Mancos Shale appears in roadcuts and on slopes between Tomichi Dome and Doyleville. The Dakota appears at the fault mentioned above and then again at Mile 176. It often caps hills north of the highway. The Morrison Formation is well exposed in roadcuts near Miles 175 and 173, lying on an irregular surface of Precambrian granite. The granite weathers into characteristic rounded boulders, and west of Parlin it is heavily iron-stained — again evidence that it is within the Colorado Mineral Belt. Also in this area you may see patches of welded tuff formed from hot volcanic ash. It is a fine tan or bluish gray rock speckled with black crystals of biotite or hornblende.

173

West of Mile 163 volcanic rocks become increasingly frequent, particularly light tawny-colored ones that, seen up close, look rather like conglomerate. These rocks came from volcanic centers to the north, in the West Elk Mountains, and are known as the West Elk Breccia, **breccia** (rhymes with betcha) being an Italian word for fragment. Rounded and angular fragments in it were thrown from volcanoes and fell here with large quantities of volcanic ash, or washed here as volcanic mudflows. The breccia lies on an old erosion surface that cuts into Morrison Shale, Precambrian granite, or Dakota Formation — whatever rock was at the surface when the volcano erupted. It often weathers into strange pinnacles and cliffs like the Palisades along the Gunnison River near Gunnison.

During the first years of its existence the town of Gunnison was a central supply point for mining activity, supplying mines on the west side of the Sawatch Range, in the West Elk Mountains to the north, and in the San Juans to the south.

u.s. 160

walsenburg — alamosa

(73 miles)

At Walsenburg, sedimentary rocks that are flat-lying or gently warped on the plains begin to come under the influence of the mountains. The town itself is close to the first sharp rise, and U.S. 160 climbs through Cretaceous sandstone that was here before the mountain uplift onto Tertiary sedimentary rocks that are younger than the mountains. The broad sloping plain of Tertiary sediments, mostly sandstone and conglomerate derived from the mountains themselves, rises westward all the way to the Sangre de Cristo Range.

Some distance north of U.S. 160, upturned Paleozoic and Mesozoic sedimentary rocks encircle the southern tip of the Wet Mountains. The narrow valley between them and the Sangre de Cristos is floored with Tertiary sediments, some of them renowned for almost perfect fossilized skeletons of *Merychippus*, four-toed ancestor of the horse. Horses were unknown in America at the time of Columbus, but the ancient lineage of their family began in the Western Hemisphere, spread via Bering Straits to Asia, and continued to develop there even as it died out in the Americas.

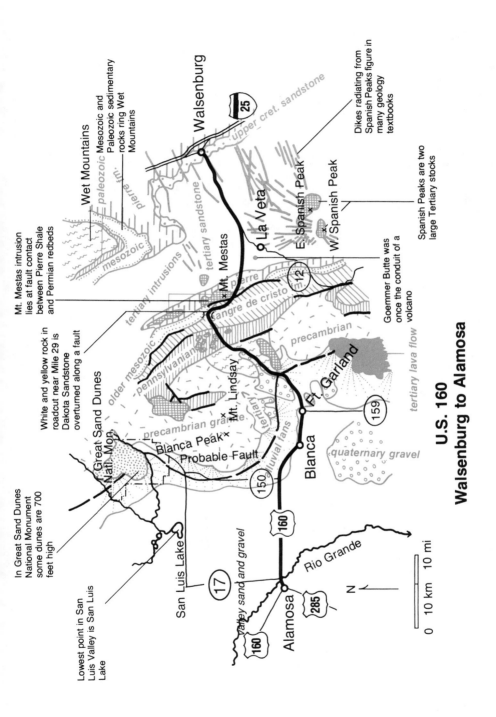

U.S. 160
Walsenburg to Alamosa

Wet Mountains

Mt. Mestas intrusion lies at fault contact between Pierre Shale and Permian redbeds

paleozoic Mesozoic and Paleozoic sedimentary rocks ring Wet Mountains

Walsenburg

upper cret. sandstone

Dikes radiating from Spanish Peaks figure in many geology textbooks

mesozoic

pierre

tertiary sandstone

La Veta

E. Spanish Peak

W. Spanish Peak

Spanish Peaks are two large Tertiary stocks

tertiary intrusions

Mt. Mestas

pierre

Goemmer Butte was once the conduit of a volcano

White and yellow rock in roadcut near Mile 29 is Dakota Sandstone overturned along a fault

sangre de cristo

precambrian

older mesozoic

pennsylvanian

Mt. Lindsay

tertiary lava flow

Great Sand Dunes

In Great Sand Dunes Natl. Mon.

In Great Sand Dunes National Monument some dunes are 700 feet high

precambrian granite

Blanca Peak

Mt.

tertiary

Ft. Garland

Blanca

quaternary gravel

Probable Fault

alluvial fans

Lowest point in San Luis Valley is San Luis Lake

San Luis Lake

valley sand and gravel

Rio Grande

Alamosa

N

0 10 km   10 mi

North and south of the highway, some of the swarm of dikes that radiate from Spanish Peaks can be seen. They cut through surrounding Cretaceous and Tertiary strata, but not through Paleozoic rocks of the Sangre de Cristo Range. JACK RATHBONE PHOTO

Spanish Peaks, visible to the south, are two large igneous intrusions that pushed their way up through Cretaceous and Tertiary sedimentary rocks after most of the rest of the Colorado Rockies were formed.

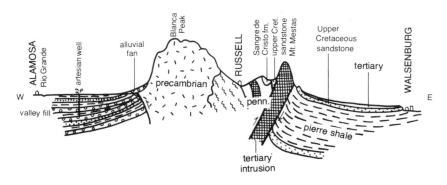

Section north of US 160, across the Sangre de Cristo Range

U.S. 160 curves northward around pyramid-like Mt. Mestas, another type of Tertiary intrusion. Geologic studies show that magma forming this mountain came up along one of the principal faults on the east side of the Sangre de Cristos, where red Permian rocks (visible west of the highway) are carried eastward over Dakota Sandstone. The intrusion is a white rock called microgranite that contains microscopic crystals of quartz, feldspar, and biotite — the same mineral trio that characterizes granite. Overlying rocks have long since washed off Mt. Mestas and the Spanish Peaks.

The fault west of Mt. Mestas is easy to find because a spring flows from it close to the road. Beyond and above the spring, dark red Permian sandstone and shale of the Sangre de Cristo Formation tint the slopes. To the right is the white microgranite of Mt. Mestas.

The Sangre de Cristo Formation extends on across La Veta Pass. Frequently shifted around by faults, it sometimes dips east, sometimes west, but it can always be recognized by its red color. It originated as an alluvial apron between Frontrangia and Uncompahgria, the two island ranges of the Ancestral Rockies.

Just beyond La Veta Pass, the summits of Blanca Peak and Mt. Lindsay can be glimpsed to the west. They are parts of a large granite mass that juts west of the main line of the range. Descending the west side of the pass the highway crosses, in order:

• A broad swale of soft red shale and sandstone of the Sangre de Cristo Formation (to Mile 276).

• A thick sequence of fossil-bearing gray Pennsylvanian shale, sandstone, and limestone (well exposed between Miles 276 and 273). Small Tertiary intrusions stick up into these rocks.

• Precambrian granite, gneiss, and schist, sometimes with yellow-green epidote on fracture surfaces (to Mile 268).

• Lavender and purple volcanic rocks (beyond Mile 268) related to the San Luis volcanic field at the south end of the San Luis Valley.

The area has had a complicated history, hard to unravel. The original faulting is probably a product of the Laramide Orogeny, but faulting seems to have continued, sporadically at least, all through Cenozoic time. When Miocene-Pliocene regional uplift brought the Rockies to their present elevations, the San Luis Valley was left behind as part of the Rio Grande Rift. Its actual floor, under 10,000 to 13,000 feet of sediments, is below sea level. A central **horst**, bounded on each side by faults, is only 5000 feet below the present surface. There is evidence that faulting is still going on, albeit at a decreasing rate. Triangular facets at the bases of mountain ridges show that some fault movement is fairly recent, as does a small fault scarp on the surface of alluvial fans in the northern part of the valley.

The huge alluvial fans at the base of Blanca Peak coalesce with others along the west side of the Sangres. They are probably similar to alluvial aprons that encircled the Ancestral Rockies in Pennsylvanian and Permian time, forming the Sangre de Cristo Formation in this area and other coarse red rocks farther north and west. Remnants of higher terraces, formed during the Ice Ages from glacial outwash, can be recognized by their even, sloping surfaces covered with rounded pebbles and cobbles.

The San Luis Valley is the only true desert in the Colorado Rockies; rainfall is less than eight inches a year. It looks perfectly flat, but the southern end of the valley, which you see here, is drained by the Rio Grande through a steep-walled gorge in volcanic flows along the Colorado-New Mexico line. The north half of the valley is a closed basin; streams entering it from the mountains never reach the Rio Grande (see U.S. 285 PONCHA SPRINGS to NEW MEXICO). Some of the valley sediments are lakebeds, deposited during episodes of volcanic damming or periods of wetter climate.

When prevailing southwest winds sweep across the valley, they carry sand from the thinly vegetated surface toward the Sangre de Cristo Range. Rising to cross low points in the range, the wind drops the sand in a sheltered corner just north of Blanca Peak, building some of America's tallest sand dunes (see GREAT SAND DUNES NATIONAL MONUMENT).

Irrigation water in the San Luis Valley comes from deep artesian wells. Some of the water from the mountains sinks into sand and gravel at the edges of the valley, and flows slowly along gravelly layers between gently sloping, impervious layers of clay or volcanic ash. There it flows downward, toward the valley, urged on by gravity and additional water from above. In wells that go down to the gravel aquifers, water rises and flows without pumping because the well head is lower than the points where the water gets trapped in the gravel. Some deep wells in this valley tap as many as seven artesian aquifers.

great sand dunes
national monument

(46-mile round trip from u.s. 160;

50-mile round trip from co. 17)

The San Luis Valley is a desert, for rainfall averages less than eight inches a year. The prevailing wind blows from the southwest, almost perpendicular to the trend of the Sangre de Cristo Range. Three low passes in this range — Mosca, Music, and Medano Passes — funnel the wind across a corner sheltered by the projecting mass of Blanca Peak, in the region of Great Sand Dunes National Monument. During spring and fall particularly strong winds pick up sand from the valley floor, drift it across the valley, and deposit it in this area.

Some of the dunes crest 700 feet above the valley floor. They consist mostly of fine, rounded grains of quartz and of volcanic rocks from the San Juans. This mixture gives the dunes a fairly dark color — darker than most dune and beach sand. Because the dark surfaces soak up solar radiation more readily than light surfaces, these dunes are sometimes far too hot for comfort. Wear thick-soled shoes if you climb on them, and don't go into the dunes at midday in summer.

The dune field is oval-shaped and about six miles across. Usually, sand drifts up the gentle windward (southwest) slopes of the dunes and then slides, rolls, and avalanches down the steep leeward (northeast) slopes. In spring, when strong storm winds sometimes swoop across the Sangre de Cristo passes from the northeast, each dune crest reverses, forming a counterdune with a gentler slope on the east and a steep one on the west.

After rains, the sweeping laminations characteristic of wind-deposited sand sometimes show up on windward surfaces — laminations that make it possible to identify dune sand even when it is hardened by time into sandstone layers.

Aerial views reveal that most of the dunes here are of the transverse type, their northwest-southeast-trending ridges lying across

179

The Great Sand Dunes are decidedly asymmetric, with gentle western windward slopes and steep eastern leeward slopes.
HALKA CHRONIC PHOTO

the path of the prevailing winds. A few crescent-shaped barchan dunes show up around the fringes of the main dune mass, and several star dunes have been identified in parts of the dune field where wind direction varies a lot. One might think that prevailing southwest winds would move the dunes slowly eastward until they bump into the mountain front. To some extent, they do, especially near the margins of the main dune field where as you can see the dunes slowly bury stands of pine trees. However, comparison of old and new aerial photos shows little net movementof the main dune ridges — they appear to have hardly changed over many decades. Perhaps the periodic storm winds from the northeast nullify the movement caused by prevailing winds from the opposite direction.

Certainly the presence of flowing creeks both north and south of the dunes regulates the position and size of the whole dune mass. The interplay between the waters of Medano Creek and the dunes is fascinating. At times, dune ridges encroach on the stream, which rarely flows all year. At other times the stream, strengthened by spring runoff, carves back the dunes, and large masses of dry sand cascade down steep lee slopes and off low cliffs undercut by the river. Much of the water of Medano Creek actually flows under rather than around the dunes, sinking into the gravel valley floor or coming to the surface in marshy springs east of the dunes.

Examine some of the stream and dune surfaces, for they have tales to tell. Stream-formed ripple marks differ in their shape and the size of their sand grains from wind-formed ripples on dry parts of the dunes. Shrinkage cracks form in the dried mud of the stream bed. On sand bars and near the stream banks tracks of dune- and forest-dwelling animals are abundant. Beetle tracks are perhaps the most common, with the slime trails of snails and the conical holes of "doodlebugs" fairly frequent also. Watch for the S-shaped groove of a lizard's dragging tail. Birds, deer, coyotes, rabbits, and humans — barefoot and otherwise — all leave characteristic trails. Grasses blown by the wind inscribe graceful circles.

Many such patterns are preserved as shifting sands drift over them. In other parts of the world, including parts of Colorado, the tracks made by long extinct animals and circles inscribed by long gone grass come to us through the mists of time, for they are known from sandstones deposited many millions of years ago in ancient dunes and near ancient rivers. Ripple marks and mudcracks too are sometimes preserved in sedimentary rocks. Perhaps some paleontologist of the distant future will puzzle over the crooked trail made by a crushed paper drinking cup blown in the wind from the National Monument picnic area!

Medano Creek, which flows only in spring or after heavy summer storms, helps to control the shape of the southern edge of the dune field.

HALKA CHRONIC
PHOTO

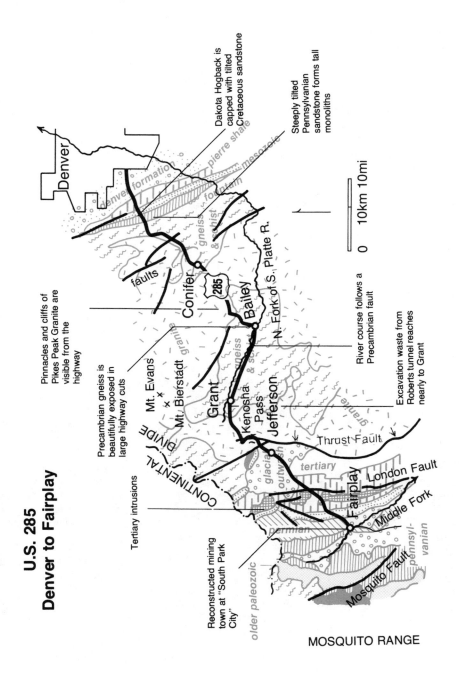

U.S. 285
Denver to Fairplay

Dakota Hogback is capped with tilted Cretaceous sandstone

Steeply tilted Pennsylvanian sandstone forms tall monoliths

Denver

pierre shale

denver formation

fountain mesozoic

gneiss & schist

faults

Conifer

Bailey

285

N. Fork of S. Platte R.

River course follows a Precambrian fault

Excavation waste from Roberts tunnel reaches nearly to Grant

Pinnacles and cliffs of Pikes Peak Granite are visible from the highway

Precambrian gneiss is beautifully exposed in large highway cuts

Mt. Evans

Mt. Bierstadt

granite

Grant

gneiss & schist

Kenosha Pass

Jefferson

CONTINENTAL DIVIDE

Thrust Fault

tertiary

granite

Tertiary intrusions

Fairplay

London Fault

Middle Fork

glacial outwash

Reconstructed mining town at "South City"

permian

pennsyl-vanian

older paleozoic

Mosquito Fault

MOSQUITO RANGE

0   10km 10mi

Ripplemarks from a Cretaceous shore mark the surface of the Dakota Sandstone. Note the rock hammer for scale.

u.s. 285
denver — fairplay
(85 miles)

As you drive toward the mountains on U.S. 285, the Tertiary pediment shows distinctly about halfway up the Front Range, an undulating surface about 8000 feet in elevation. By the time you get to Conifer you will be up on this surface.

Between Denver and the mountains the highway crosses Tertiary and Cretaceous rocks, rarely exposed in the suburban area. However, at the Dakota Hogback Cretaceous and Jurassic sediments are tilted steeply by the uplift of the Rockies. The hogback is formed of resistant Dakota Sandstone laid down as a Cretaceous seashore deposit. Petrified ripple marks are visible in the highway cut; dinosaur tracks have been found here as well.

Between the hogback and mountain front the highway quickly crosses successively older strata: Jurassic purple and green shale of the Morrison Formation, Triassic and Permian red sandstone and limestone, and Pennsylvanian sandstone that stands up as tall pink monoliths near the mountain front. Once flat-lying, all these rocks were tilted into their present position by mountain uplift. Behind them, separated by more than a billion years of time, are Precambrian rocks. Older Paleozoic strata, though deposited here, were washed away when this area was part of the highland of the Ancestral Rockies, in Pennsylvanian time.

Precambrian gneiss and schist form most of the mountain core along this route. They vary in color from light gray to almost black, depending on the proportion of black minerals (hornblende and biotite) and light minerals (quartz and feldspar). Their radioactivity clocks tell us they are 1750 million years old, and they seem to be relics of very old mountains eroded completely away by the end of Precambrian time. Occasional pegmatite dikes contain large crystals of biotite, quartz, and feldspar.

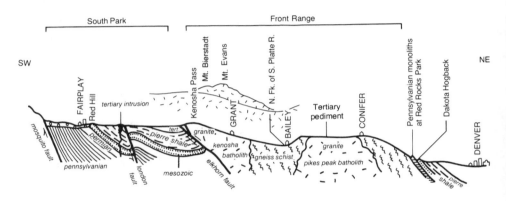

Section along U.S. 285 – Denver to Fairplay

Most of the many faults that have been mapped in these Precambrian rocks trend northwest-southeast rather than parallel to the north-south trend of today's Front Range. Careful study of these faults, and of the dikes and the "grain" or fabric of the rock itself, has led to a belief that the trend of the present mountains is not related to the trend of the mountains that were here in Precambrian time.

A few miles into the mountains (Mile 241) the highway crosses a bulbous northern arm of the Pikes Peak batholith, an immense body of coarse pink granite intruded into the metamorphic rocks about a billion years ago. The granite is easy to recognize, for it is lighter in color and more uniform in texture than the metamorphic rocks, and usually has large, coarse crystals. In some places it forms knobs, massive cliffs, and pinnacles of monumental rounded blocks. It often decomposes into pink or salmon-colored sand.

184

At Bailey the highway, once again in the schist-gneiss complex, enters the canyon of the North Fork of the South Platte River. This canyon parallels the ancient faults. The course of the river here is controlled by one of them, with the river cutting through broken and weaker rocks to avoid more solid rocks nearby. The highway passes just south of Mt. Evans (14,269 feet) and Mt. Bierstadt (14,060 feet), whose summits are composed of granite similar to that near Conifer.

Glacial and river gravels (not shown on the map) form occasional flat terraces along the river. They date from the Pleistocene Ice Ages, when the river carried much more water and sand and rock than it does now. The highway crosses a hilly glacial moraine near Mile 213.

Just west of Grant is the portal of the 23-mile-long Harold D. Roberts Tunnel, carrying water from Dillon Reservoir west of the Continental Divide to the South Platte east of the divide, for Denver's use. Colorado's large eastern-slope population centers depend increasingly on western slope waters.

Several miles west of the tunnel portal, the highway once more crosses into granite, this time the Kenosha batholith. Again the contact is irregular, but here isolated granite fingers seem to float among the flow lines of the metamorphic rock. The granite is often quite badly fractured, again suggesting that there may be faults here.

There are lots of beaver dams along this highway. Active little geologic agents, beavers create ponds that in time fill in and become meadows; their dams decrease erosion and often hold back flood waters. Hummocky areas near the pass are glacial moraines.

The overlook just beyond Kenosha Pass (Mile 203) offers a splendid view of South Park, one of three large intermontane valleys in Colorado, a faulted syncline between the Front Range and the Sawatch uplift. The east side of South Park is edged by the Elkhorn thrust fault, with Front Range Precambrian rocks pushed up and over westward onto Cretaceous and early Tertiary rocks, probably in response to the decrease in confining pressure as the Front Range rose. On the west side of South Park, in the Mosquito Range, Paleozoic formations rise to the crest of the range. Halfway along the west margin of South Park, behind the long ridge of the Dakota Hogback, you can see Buffalo Peaks, a mountain formed of Tertiary volcanic rocks deposited in an ancient valley. Looking back to the northeast you can see Mt. Evans and Mt. Biestadt.

Much of South Park is underlain by Paleozoic and Mesozoic sedimentary rocks, which are mostly concealed by a skimpy veneer of gravel. These rocks dip east, and become older and older westward on the flanks of the Mosquito Range, in exactly the same sort of pattern as on the east side of the Front Range. Descending from the pass, the highway crosses them in this order:

• A wide valley underlain mostly by Cretaceous Pierre Shale.

• At Mile 190 a large ridge-forming igneous dike.

• Red Hill, a long hogback of tilted Cretaceous and Jurassic sediments, the Dakota and Morrison Formations, similar to the Dakota Hogback near Denver. Red rocks are part of the Morrison Formation.

• Another valley, this one underlain by Permian and Pennsylvanian redbeds equivalent to the Fountain Formation near Denver. These rocks, covered at the edge of the valley by glacial deposits, extend up onto the shoulder of the Mosquito Range.

Fairplay, established in 1859, was a lively and by all reports often unruly gold camp. The first gold was obtained by panning river gravels, or by shoveling them into crude homemade rockers. Then for years gold was obtained from valley-floor gravels by washing them with high-pressure hoses and running the streams of rock and sand through sluiceboxes, where the gold was caught in corrugated "riffles" lined with gunny-sacking. In 1922 a gold dredge was built on the Middle Fork of the South Platte to process river gravels. A similar dredge still floats on its pond a few miles south of Fairplay; it ceased operations in 1952. The giant arcuate heaps of dredge tailings can still be seen near Fairplay — no environmental consciousness then!

The South Park Ranger District office distributes brochures for two self-guided driving tours of mining areas near Fairplay. South Park City is an interesting outdoor museum containing many original buildings and artifacts dating from the gold camp days.

The gold dredge near Fairplay was built in 1941, and operated until 1952. It dug gravel 70 feet below water level, collapsing the 35-foot bank above the water level. So it really went through 105 feet of gravel as it moved slowly along, digging the bank ahead and dumping the tailings behind. The gravel was run through sorters and sluice boxes right on the dredge, which in 11 years recovered 115,000 ounces (more than 3½ tons!) of gold from about 33 million cubic yards of gravel.

u.s. 285

fairplay — poncha springs

57 miles)

Where U.S. 285 crosses the Middle Fork of the South Platte near Fairplay, tailings piles from old dredging edge the stream. A short distance southeast of Fairplay a large old gold dredge floats on its self-made pond. It is larger than it looks, for the pond is 35 feet below the terrace surface. Beyond it arch the mounds of rounded cobbles and boulders deposited by its great conveyor buckets.

The highway rises almost immediately onto a terrace of stream-deposited glacial gravel, with occasional exposures of red Permian Maroon Formation. For an interesting side trip, taking you to the older Paleozoic rocks of the Mosquito Range, drive up Fourmile Creek to Horseshoe Mountain Cirque. (Self-guided tour brochures are available at the Fairplay Ranger Station.) There, above the old mining town of Leavick, glacial erosion has exposed sedimentary layers that range in age from Cambrian to Pennsylvanian.

187

U.S. 285
Fairplay to Poncha Springs

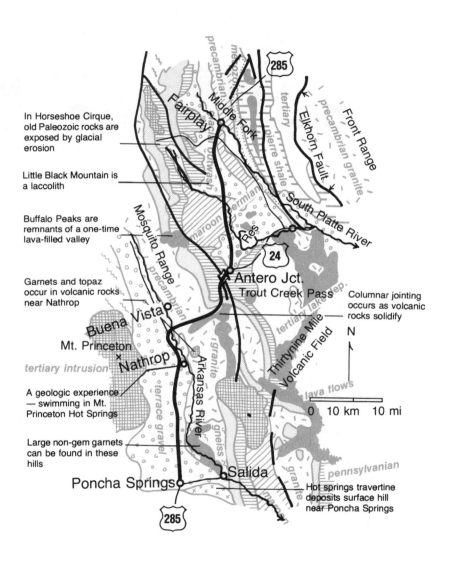

In Horseshoe Cirque, old Paleozoic rocks are exposed by glacial erosion

Little Black Mountain is a laccolith

Buffalo Peaks are remnants of a one-time lava-filled valley

Garnets and topaz occur in volcanic rocks near Nathrop

A geologic experience — swimming in Mt. Princeton Hot Springs

Large non-gem garnets can be found in these hills

Columnar jointing occurs as volcanic rocks solidify

Hot springs travertine deposits surface hill near Poncha Springs

N

0   10 km   10 mi

Fairplay
Middle Fork
precambrian
mesozoic
tertiary
pierre shale
permian
maroon
outwash
Front Range
precambrian granite
Elkhorn Fault
South Platte River
Res.
24
285
Mosquito Range
precambrian
Antero Jct.
Trout Creek Pass
tertiary lake dep.
Thirtynine Mile Volcanic Field
Buena Vista
Mt. Princeton
tertiary intrusion
Nathrop
granite
Arkansas River
terrace gravel
lava flows
gneiss
Salida
pennsylvanian
granite
maroon
Poncha Springs
285

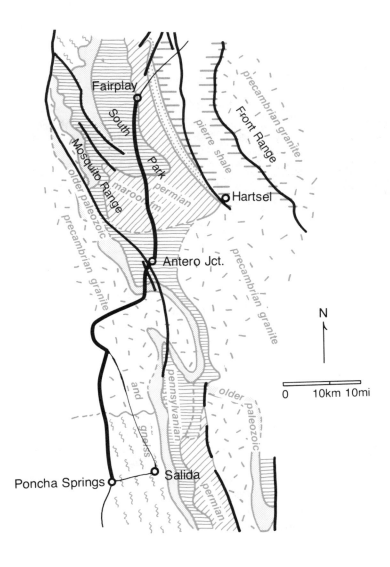

The map of this section of U.S. 285 (opposite) looks hopelessly complex.
But if all Tertiary and Quaternary deposits are omitted, a fairly simple
pattern emerges – one of more or less north-south bands of Paleozoic
and Mesozoic sedimentary rocks somewhat faulted and folded and
caught between two uplifted blocks of Precambrian crystalline rocks.
Youngest Paleozoic rocks are on the east side of South Park, oldest on
the west, indicating that this is but the eastern half of the huge
Mosquito Range-Sawatch Range faulted anticline.

West of Fairplay, in the Mosquito Range, glacial erosion has bared Paleozoic sedimentary rocks in Horseshoe Cirque (left of center). A glacial lake occupies the cirque, below magnificent rock glaciers of fallen blocks of rock lubricated with water and ice. The slopes of Mt. Sheridan and Mt. Sherman to the north (right) are peppered with old gold mines, from which ore used to be carried down in buckets on a cable lift.

T.S. LOVERING PHOTO, COURTESY OF USGS

Buffalo Peaks, the large, dark, double-summited mountain southwest of Fairplay, is a remnant of the many thick layers of andesite lava and volcanic ash that filled a Tertiary valley (**Andesite** is a gray, fine-grained igneous rock composed chiefly of feldspar.) Remnants of related flows also cap several small buttes nearby. But Little Black Mountain, the round-topped mountain between Buffalo Peak and Mile 178, is a small laccolith, in which molten magma domed up overlying sedimentary strata but did not escape to the surface.

Pikes Peak is now in the distance to the southeast, beyond humpy hills of the Thirtynine-Mile volcanic field. A hogback ridge of Paleozoic sedimentary rocks that edges South Park on the west appears straight ahead, with the distant summits of the Sangre de Cristo Range on the southern skyline beyond it.

As the highway turns toward Trout Creek Pass, you'll see Pennsylvanian black shale and thin gray limestone in roadcuts. These rocks predate uplift of the Ancestral Rocky Mountains. Since this area was between the two ranges of the Ancestral Rockies, and so was not lifted

190

and exposed to erosion late in Paleozoic time, older Paleozoic layers were not eroded away.

Trout Creek Pass is right on a fault, one of several that run along the Mosquito Range. West of the pass the rocks are down-dropped. At Mile 223-224 the highway is in a small syncline of Paleozoic rocks. The sequence here is:

• Pennsylvanian black marine shales and thin limestones, with some coaly layers and fossil plants suggesting a swamp environment.

• Mississippian Leadville Limestone, gray and massive, forming the skyline ridges above both sides of the synclinal valley.

• Devonian light brownish limestone and fine white sandstone, forming slopes and ledges but usually poorly exposed. Horn corals and flat tabulate corals occur in this rock.

• An Ordovician cliff-slope-cliff sequence of two massive limestone units sandwiching a sandstone-shale unit, visible at the narrows just above the bridge over Trout Creek.

Fossils are fairly common in these rocks, particularly in the Leadville Limestone and the upper of the Ordovician cliffs, where horn corals are easy to find.

The highway descends toward the Arkansas Valley through rugged country studded with huge rounded granite boulders. The granite is cut by white pegmatite dikes containing many unusual min-

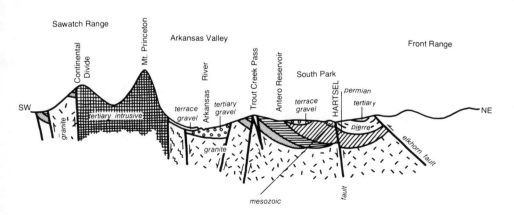

Section across South Park and the Sawatch Range at Trout Creek Pass.

191

erals such as beryl, uxenite, tantalite, and xenotime, not often found in such abundance. Weathering of granite follows a typical pattern, with mica grains decomposing into clay minerals, popping off quartz grains and sometimes whole curving sheets of the rock surface. The rounded boulders you see here result, and sand around the base of the boulders is feldspar and quartz sand, with grains the same size as the crystals of the parent rock.

The granite here is about 1750 million years old. It is part of the core of the Sawatch uplift, a broad faulted anticline sliced in two by the Rio Grande Rift, a fault valley occupied by the Arkansas River. The Mosquito Range is the eastern half of the Sawatch uplift anticline, the Sawatch Range the western half. High peaks of the southern Sawatch Range are visible as you descend into the Arkansas Valley. Though the northern part of this range is composed of Precambrian rock, Mt. Princeton directly facing you, and Mt. Antero and Mt. Shavano farther south, are parts of a much younger intrusion that melted and pushed in here as molten magma early in Tertiary time.

Many Sawatch Range peaks, including the three listed above, are Fourteeners, with summits over 14,000 feet in elevation. The Arkansas Valley here is just below 7800 feet.

On all these high peaks, cirques can be seen at high elevations. Glaciers have left their signatures — in the form of U-shaped valley profiles — between the peaks as well. Terminal moraines cross these tributary valleys just about at the edge of the Arkansas Valley, at almost exactly 8,000 feet elevation — the magic (but not inviolate) number for the lower limit of glaciation in Colorado. Along the base of the Sawatch Range a row of alluvial fans containing stratified and sorted glacial outwash has coalesced into an almost continuous apron. No such huge fans have developed along the east side of the valley, as the Mosquito Range was not glaciated here in its lower, southern part.

The eroded white patch on the southeast flank of Mt. Princeton, visible from Mile 143-141, is kaolinite, a soft chalky rock formed by leaching by hot water rising along the faults that edge the valley. The powdery white rock gives the name to Chalk Creek, though it has a composition quite different from that of real chalk. There are still hot springs here, and at Princeton Hot Springs there is a spa and swimming pool. Have a geological swim!

The Arkansas River follows a long, straight path right down the Precambrian core of the Sawatch uplift, along the line of this northern part of the Rio Grande Rift. Apparently it once continued far southward and into the Rio Grande, for an ancient channel has been discovered, now filled with volcanic rocks, that would have taken it directly into the San Luis Valley south of Poncha Pass.

Terraces along the Arkansas contain gravel and boulders of glacial outwash brought by streams from glacial moraines near the head of the valley.

There are also hot springs at Poncha Springs, located fairly high up on the slope south of the town. They flow only about 15 gallons a minute, but they are quite hot, 150°F. Most of the water is piped through an insulated pipe to the big indoor swimming pool at Salida.

u.s. 285

poncha springs — new mexico

(126 miles)

South of Poncha Springs, U.S. 285 crosses Poncha Pass, the saddle between the Sawatch and Sangre de Cristo Ranges. For much of the way over the pass the highway is bordered by Precambrian hornblende gneiss, recognizable by its dark gray color and abundant white pegmatite dikes. There is also schist here, and some striking black and white banded gneiss. Much of the Precambrian rock is **hydrothermally** altered by percolating hot water, which has changed most of the feldspar minerals into a white clay mineral called **kaolin**.

At Mile 122, Tertiary volcanic rocks begin to show up as patches of purplish lava flows near the road. Farther southwest, similar dull red or purple flows have piled up one on top of another for thousands of feet to form the magnificent San Juan Mountains, by far the largest volcanic region in Colorado (see Chapter V). The highway skirts the east side of these mountains between Saguache and Alamosa, and even slices through a flow that reaches long fingers out into the San Luis Valley.

Poncha Pass marks the divide between Arkansas River drainage to the north and Rio Grande watershed to the south. But recent research shows that this wasn't always the case. In Miocene and Pliocene time the Arkansas flowed southward into the San Luis Valley and the Rio Grande, through a valley or channel just west of Poncha Pass. Then

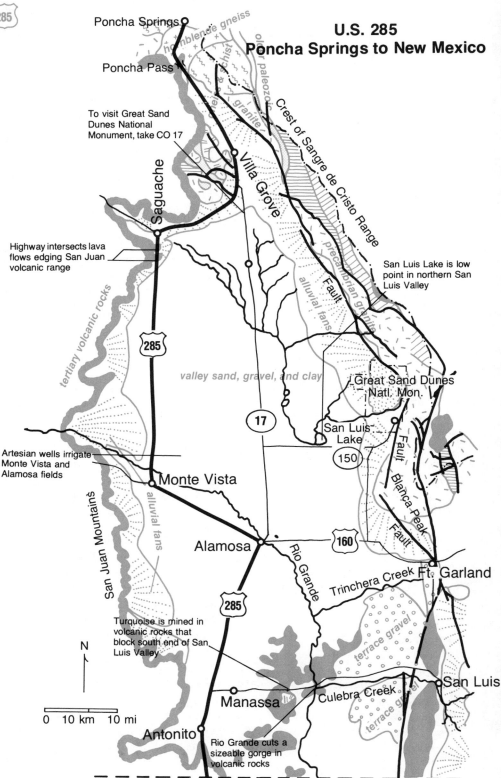

U.S. 285
Poncha Springs to New Mexico

Poncha Springs

Poncha Pass

To visit Great Sand Dunes National Monument, take CO 17

hornblende gneiss

older paleozoic

schist & slate

granite

Crest of Sangre de Cristo Range

precambrian granite

Fault

Saguache

Villa Grove

San Luis Lake is low point in northern San Luis Valley

Highway intersects lava flows edging San Juan volcanic range

tertiary volcanic rocks

alluvial fans

285

valley sand, gravel, and clay

Great Sand Dunes Natl. Mon.

17

San Luis Lake

Fault

Blanca Peak

150

Artesian wells irrigate Monte Vista and Alamosa fields

Monte Vista

alluvial fans

San Juan Mountains

Alamosa

Rio Grande

160

Fault

Ft. Garland

Trinchera Creek

285

Turquoise is mined in volcanic rocks that block south end of San Luis Valley

N

terrace gravel

San Luis

Manassa

Culebra Creek

terrace gravel

0   10 km   10 mi

Antonito

Rio Grande cuts a sizeable gorge in volcanic rocks

uplift and volcanic action closed that valley and forced the river into a new route out of the mountains.

The San Luis Valley, bordered on the east by the Sangre de Cristo Mountains and on the west by the San Juans, extends from Poncha Pass almost to the New Mexico border. As large as Connecticut, and far deeper geologically than it looks, it has been filled with layer after layer of gravel, sand, clay, lava, and volcanic ash. Its real floor is 10,000 to 13,000 feet below the present surface.

The north half of the valley makes a closed basin that has no outlet. Streams entering it disappear into porous sands of alluvial fans and

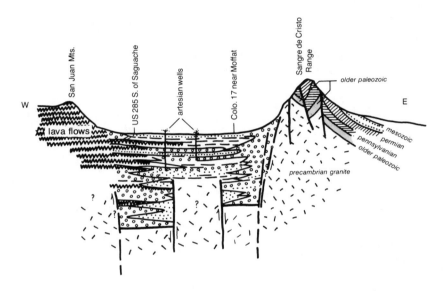

Section across northern part of San Luis Valley

valley floor, or find their way into a natural sump at San Luis Lake, near Great Sand Dunes National Monument. If the climate were less desert-like, as it has been in the past, a lake would fill the north half of the basin.

Prevailing west and southwest winds frequently blow sand from the valley floor to the east side of the valley, where it accumulates between Blanca Peak and the northern part of the Sangres in a sea of towering dunes that rise 700 feet above the valley floor (see GREAT SAND DUNES NATIONAL MONUMENT).

The south part of San Luis Valley, separated from the northern closed basin by an almost imperceptible rise near Monte Vista and

Alamosa, drains into the Rio Grande. This river and its tributaries rise high in the San Juan Mountains, where heavy winter snowpack provides the abundant water necessary to erode the mountains and to carry the debris into the valley. Deep-lying deposits in the valley are Paleocene, showing that the valley became established as the Rockies formed. Deposition continued through Miocene-Pliocene uplift and on through Quaternary time. Layers of volcanic ash and lava from the San Juan Mountains alternate with layers of sedimentary gravel, sand, and clay.

Many theories have been advanced as to why an area like the Rio Grande Rift, of which the San Luis Valley is a part, should have formed. One theory proposes that when a portion of the curving earth's crust is lifted, its total east-west dimension has to increase. So, under tension, the crust cracks, and a slender block between two cracks (or faults) drops or remains relatively stationary while blocks on both sides rise. Another theory suggests that clockwise rotation of the plateau area of Arizona, Utah, Colorado, and New Mexico pulls the edges of the valley apart. In any case, the rifting seems to be due to extension, or drawing apart, of parts of the earth's crust, so the word "rift" is appropriate.

The steep western face and high jagged crest of the Sangre de Cristo Range marks the line of a large fault zone bordering the east side of the valley. Though movement on this fault zone began during the Laramide Orogeny, some — perhaps most — of it took place in Miocene-Pliocene time, when the Rio Grande rift did not rise during regional uplift. There is evidence that sporadic movement is still going on: notice the triangular facets at the bases of mountain ridges — facets touched only lightly by erosion. Even in the last few thousand years a fault scarp has developed in alluvial fans along the base of the mountains. On the east side of the range, uplift was accomplished by both faulting and tilting. Paleozoic sedimentary rocks extend clear up the east side to the crest of the range; their cut edges make up part of the western face.

Groundwater geology of the San Luis Valley is particularly interesting. Large canals built around 1880 used to bring irrigation water from Rio Grande tributaries into the closed non-draining north half of the valley. Farming was good for a while, but eventually the water table (the surface of the groundwater) rose nearly to the surface, and farm land became waterlogged and had to be abandoned. So farmers moved south to the drained part of the valley, near Monte Vista and Alamosa. Unfortunately, starting in the 1930s, these farms suffered from a long drought, with runoff below normal for 20 to 30 years. Desperate farmers turned to subsurface water, drilling

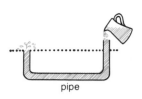 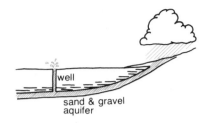

pipe

well

sand & gravel
aquifer

In an artesian system, the point of entry is higher than the wellhead, so well water rises without pumping. Impervious shale or clay layers prevent the water from rising except where wells are drilled.

wells deeper and deeper into the valley floor. This time they were luckier. Because of the trough-shaped configuration of the valley and the alternating porous gravel and impervious clay and ash and lava layers that fill it, their wells were artesian — water rose and flowed from them without pumping. There are more than 7000 flowing wells in the valley now. And the use of well water lowers the water table, so the northern half of the valley is again becoming arable. But unrestricted use of artesian water will eventually deplete the water supply.

Blanca Peak juts westward from the otherwise linear Sangre de Cristo Range as an immense extension of the Precambrian rock that

A 25-foot scarp in an alluvial fan shows how recently the Sangre de Cristos have shifted. The car is below and west (left) of the fault line; the photograph is taken from the upthrown side of the scarp, which is only slightly modified by erosion. Across the the San Luis Valley rise the volcanic peaks of the San Juans.

G.R. SCOTT PHOTO, COURTESY OF USGS

underlies the mountains. The great fault system that edges the range seems for some reason to bulge westward here, though its exact location, somewhere under the alluvial fans, isn't known.

Near Monte Vista the highway crosses the Rio Grande, not as grande now as it used to be. Water is taken from it all the way from its source in the high peaks of the San Juans to its mouth in the Gulf of Mexico. The river turns south near Alamosa and plunges into a narrow black-walled canyon through the volcanic rocks that dam the south end of the San Luis Valley. Looking east from Mile 12 you can see the volcanic hills near Manassa, some of them capped by remnants of flat-lying lava flows. Here, turquoise is mined, a product of hydrothermal alteration of copper minerals in the volcanic rock.

From Antonito a little narrow-gauge railroad winds its way (summer only) up Pine Canyon and across Cumbres Pass to Chama, New Mexico. The trip is an interesting one geologically, for rocks range in age and composition from Precambrian crystalline ones to Pleistocene volcanic and glacial deposits. Several guidebooks available in the "depot" outline the geology along the route.

The dome-shaped mountain to the south, San Antonio Peak, is one of a cluster of small volcanoes which, with their related lava flows, dam the San Luis valley, preventing escape of groundwater and, for a time in the past, lake water.

u.s. 287
wyoming — denver

(95 miles)

(see map for I-25

wyoming — denver)

One of the most interesting routes by which to enter Colorado is U.S. 287, which crosses the Wyoming line among hills of pink granite, most of it gray-surfaced with lichens. This rock is part of the core of the Front Range, one lobe of a very old mass of igneous rock called the Sherman Batholith, intruded during a mountain-building episode in Precambrian time. Radioactivity age determinations indicate that it is at least 1.4 billion years old. And what a history it has had since its creation!

• Worn down to sea level during the long erosional interval at the end of Precambrian time.

• Covered with layer after layer of Lower Paleozoic marine sediments.

• Uplifted during the Pennsylvanian Period to form the Ancestral Rockies.

• Stripped and laid bare once more by erosion.

• Covered again, with more than 15,000 feet of Pennsylvanian, Permian, and Mesozoic continental and marine sediments.

• Lifted another time during Laramide mountain-building.

• Cleaned off anew by another cycle of erosion.

• Lifted about 5000 feet more to its present elevation.

The granite is composed of quartz, mica, and pink feldspar grains of almost uniform size. Everywhere it is jointed in a more or less cubic pattern, each set of joints reflecting a different direction of stress to which the rock has at some time been exposed. Where it is weathered it rounds to titanic boulders or decomposes to coarse pink sand. It extends into the Sherman Mountains (part of the Laramie Range) in Wyoming, as well as south from the state line along the highway for about ten miles.

Two miles south of Virginia Dale, the highway rises to a relatively smooth, bare granite surface that is probably very ancient, a pediment or peneplain created during erosion of the Ancestral Rockies — a 270 million-year-old roadbed! Nearby, deep red Pennsylvanian sandstone and shale of the Fountain Formation lie directly on the granite surface. The contact seems sharp when seen from a distance, but the surface of the granite actually is deeply weathered **under** the Pennsylvanian sediments.

Older Paleozoic rocks — Cambrian to Mississippian — almost surely once covered the granite here. As the Ancestral Rocky Mountains rose, however, these were stripped away. Along the mountain front, layers of sand and silt were deposited, just as they are being deposited along the mountain fronts now. The Pennsylvanian layers hardened into the Fountain Formation, which edged the ancient mountains.

Continuing southward, you will begin to see other rock layers forming cuestas and hogbacks east of the Fountain Formation. Resistant light brick red Pennsylvanian or Permian sandstones often cap these features. Notice how one hogback curves around a prong of granite before paralleling the highway for several miles. When the highway at last passes through it, it is confronted with another prominent hogback, this one known to geologists and non-geologists alike as the Dakota Hogback because it is capped by Cretaceous rock, the Dakota Sandstone. For ten miles the highway runs between the two hogbacks, the Pennsylvanian one on the west and the Dakota on the east. The valley between the two is eroded in soft Triassic and Jurassic shales and mudstones.

As you drive along the mountain front south of Fort Collins you will continue to see the scalloped crest of the Dakota Hogback close to the mountains. Horsetooth and Carter reservoirs use it as a containing eastern wall.

As you leave the mountains, you can see more of the range itself. The high part west of Fort Collins and Loveland is in Rocky Mountain National Park. This includes a prominent peak to the southwest, Longs Peak, at 14,255 feet a Fourteener in good standing. Although

200

*Tilted Paleozoic and Mesozoic strata in the hogbacks fringing
the mountains are all that remain of the blanket of
sedimentary rock that once extended across what was to
become the Precambrian core of the Front Range.*

you can't see it well from this angle, the summit of Longs Peak is flat, and about the size of a city block. Some geologists think the flat summit may be an exhumed remnant of an erosion surface dating from the Lipalian Interval at the end of Precambrian time. A similar surface can be recognized in oilwell cores from 13,000 feet underground near Denver, giving us a measure of vertical displacement along the faults that edge the Front Range.

Between Fort Collins and Denver, U.S. 287 passes over Cretaceous shale and sandstone, but good exposures are rare. Soil derived from the shale can best be recognized by its fertility. Soils developed on the sandstone are less fertile and usually good only for grazing land. A few miles south of Fort Collins, where Fossil Creek runs through the shale, there used to be good places to find well preserved fossil clams and ammonites; now the area is in high demand for homesites.

Quaternary alluvium fills river valleys here, and low Pleistocene (Ice-Age) terraces are evident along some of them. Parts of Fort Collins and Longmont are on two of these terraces, which were established as river floodplains during the Ice Age and then were cut into by less heavily loaded streams after the melting of the glaciers.

201

Colorado 82
Glenwood Springs to Aspen

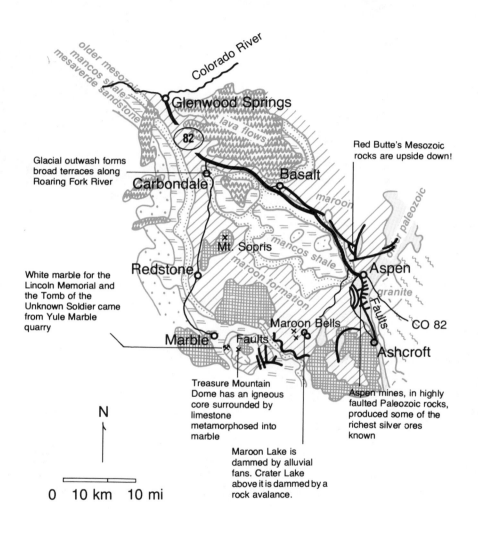

older mesozoic

mancos shale

mesaverde sandstone

Colorado River

Glenwood Springs

82

lava flows

Glacial outwash forms broad terraces along Roaring Fork River

Carbondale

Basalt

Red Butte's Mesozoic rocks are upside down!

maroon

older paleozoic

× Mt. Sopris

mancos shale

Redstone

maroon formation

Aspen

granite

White marble for the Lincoln Memorial and the Tomb of the Unknown Soldier came from Yule Marble quarry

Maroon Bells

CO 82

Marble

Faults

Faults

Ashcroft

Treasure Mountain Dome has an igneous core surrounded by limestone metamorphosed into marble

Aspen mines, in highly faulted Paleozoic rocks, produced some of the richest silver ores known

Maroon Lake is dammed by alluvial fans. Crater Lake above it is dammed by a rock avalance.

N

0 10 km 10 mi

co. 82
glenwood springs — aspen
(42 miles)

The town of Glenwood Springs has little elbow room in the narrow canyon of the Colorado River, and not much more in the lower valley of Roaring Fork River. Expansion means building on steep canyon walls, not an easy matter and fraught with geologic risks. Mudslides and earthflows, in which slopes saturated with water break apart and move downhill, have repeatedly destroyed and damaged buildings and endangered human life.

South of Glenwood Springs along Co. 82, Permian rocks of the Maroon Formation outcrop in slopes on the east side of the valley of Roaring Fork River. You can't miss these redbeds, formed of sand and clay and gravel washed from the Ancestral Rocky Mountain range of Uncompahgria during Pennsylvanian and Permian time.

The floor of the valley contains thick deposits of river gravel, most of it glacial outwash, for the upper parts of this valley and its tributaries once contained sizable glaciers. Terraces add graceful horizontal lines to the valley floor. Each terrace level represents a period of stability in the history of the river — stability of gradient, of water quantity, and of stream load or gravel and sand supply. The highest terraces are the oldest (see I-76 JULESBURG — FORT MORGAN). As you progress toward Aspen, you will see as many as three terrace levels, probably representing three glacial episodes in Colorado. The terraces contain rounded cobbles of many colors, including red rocks derived from the Maroon Formation, black basalt from volcanic hills to the north, and pale gray granite from the heart of the Sawatch Range.

On the west side of the valley, soft yellow Tertiary shale and sandstone appear in steeply eroded slopes. Above and beyond these outcrops is a sequence of Late Paleozoic and Mesozoic sandstone and shale, all dipping steeply southwest, culminating in a hogback ridge of Mesaverde Formation, coarse Cretaceous sandstone, which defines the west edge of the Southern Rockies. Formations between the Mesaverde and the Tertiary outcrops are thickly forested. The successive ridges they form are, however, visible from the road up to the Colorado Mountain College West Campus, which lies on some of the volcanic rocks that cap the red hills to the east.

Mt. Sopris, rising to the south, is a large Tertiary stock, an intrusion of crystalline igneous rock called **quartz monzonite**, containing two types of feldspar, a little quartz, and a few dark minerals. A well developed rock glacier of large angular boulders creeps down the cirque valley below the nearer summit. Like true glaciers, rock glaciers are important geologic agents, reshaping their valleys by grinding rock with rock, and building up moraine-like ridges at their lower ends. They are common in high, steep, glaciated areas where large quantities of rock are loosened by frost and lubricated by ice.

Near Carbondale, Colorado 133 branches off to the right. A side trip south along this partly unpaved route will take you eventually to the interesting old town of Marble. The road follows the canyon of the Crystal River, between red slopes formed of Maroon Formation sandstone and shale (the rock that gave a name to the town of Redstone), and turns with the river into a glaciated valley carved in younger rocks, mostly soft, drab-colored Cretaceous Mancos Shale. Valley slopes here, oversteepened by glaciation, are prone to turn into mud flows, as you can see around Marble. The town was once virtually destroyed, but sprang up again just west of its original site. Recent attempts to develop the area for skiing have been abandoned because of the instability of the slopes above the town.

Mt. Sopris rock glaciers continue the erosional work of Ice Age glaciers. Heavily loaded streams supplied by glaciers formed the foreground terraces. The mountain itself is a stock, a large intrusion of igneous rock. JACK RATHBONE PHOTO

Yule Marble quarries, source of the exquisite white marble of the Lincoln Memorial and the Tomb of the Unknown Soldier, are on the flanks of Whitehouse Mountain about two miles southeast of Marble. Snowy white and beautifully free from flaws, the marble formed by metamorphism of Leadville Limestone in an area domed up and intruded by an igneous stock. Remote from markets and plagued by transportation problems, not the least of which were caused by the slide-prone Cretaceous and Pennsylvanian shales, the quarries closed in 1940.

The town of Basalt is surrounded by boulders of the rock for which it was named, fallen from the lava-capped mesa north of the town. This gray to black volcanic rock, often full of bubble holes formed when the rock was molten, was so fluid as a magma that it spread into broad almost horizontal sheets. It probably did not come from a volcano in the traditional sense, but welled up from fissures and cracks in the ground, quietly flowing outward onto a nearly level surface. The basalt flow covers a sizeable area to the north and was once continuous with the lava cap to the west, now separated from it by the Roaring Fork River valley.

Near Basalt, Triassic redbeds appear on hills on either side of the road. They are more or less continuous with the Pennsylvanian-Permian redbeds you have already seen, and show that in Triassic time Uncompahgria was still high enough to provide a source of sediments. Above the Triassic redbeds, Jurassic Morrison Shale often forms mounded summits. It was laid down in widespread swampy deltas in a region devoid of mountains and showing no evidence of remains of the Ancestral Rocky Mountain highlands.

With the many faults in this area, sedimentary rocks may be strangely tilted this way and that. Near Mile 30, watch for a small anticline caused by flowage of gypsum. Gypsum tends to flow under pressure like silly putty (only more slowly), and here it has bulged into a plug-like mass under the present surface. Because it is lighter than overlying rocks, the gypsum plug seeks to reach the surface, as a drop of oil seeks to rise through water. So far, it has succeeded only in doming up some of the overlying rocks.

Sometimes rock layers are overturned by movement along a fault. Mesozoic sedimentary layers, for instance, are completely upside-down on Red Butte, to the left at Mile 40. Triassic redbeds at the top give the butte its red color. They are underlain by younger pink Jurassic sandstone and varicolored Morrison shales, and these are underlain by still younger Cretaceous sandstone and shale.

205

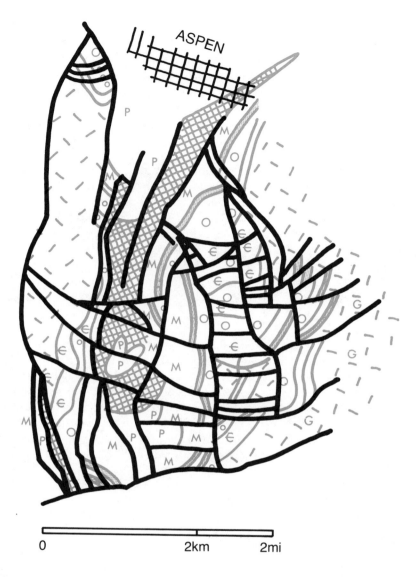

ASPEN

One can't help but be impressed by the complex faulting on the mountain slope south of Aspen. Several hundred faults have been mapped. All these structural complexities, now below Aspen's ski lifts, have been smoothed over and planted with grass.

P   – Pennsylvanian
M   – Mississippian
O   – Ordovician
Є   – Cambrian
G   – Precambrian granite

Near Mile 42 the road to Maroon Bells branches off to the right. This side trip you are probably intending to take anyway. An itinerary follows this section.

At Aspen, older Paleozoic rocks reach the surface. They are much cut by faults — many more than show on the map. The most prominent Paleozoic formation is the gray, massive Leadville Limestone, a Mississippian rock which forms the slopes of West Aspen Mountain, straight ahead at Mile 42. Many of the hundreds of faults in this area run across the slopes under the ski lifts, slopes which in Aspen's mining heyday (1879-1893) were dotted with mines. Stairstep faults bring mineral-rich Paleozoic rocks, particularly Cambrian, Ordovician, and Mississippian strata, to the surface again and again. Ores were deposited when concentrated solutions from intruding granitic magma worked their way into the thousands of cracks and fissures formed during the Laramide Orogeny.

Here as in most mining areas we are faced with a geological whodunit: Are faults unusually abundant in this area, and does their very abundance account for the mineral enrichment? Or do there simply **seem** to be more faults here where the area has been gone over inch by inch and foot by foot, above ground and below, in the search for elusive riches?

Aspen is above the lower limit of glaciation: the oldest moraines extend two to three miles below the town. Many glacial features can be seen by driving up Castle Creek to Ashcroft. You will be following the Castle Creek Fault, one of the largest in the area. Several moraines appear in the first few miles, and the U-shaped valley, bordered by lateral moraines that support thick stands of aspen and evergreens, is one of the loveliest in Colorado.

The awesome scenery at Maroon Bells results from a succession of geologic events: the rise of the Ancestral Rockies in Pennsylvanian time, the deposition of iron-bearing sediments around the island ranges of the uplift, later uplift and almost simultaneous metamorphism of the red rocks, further uplift in Miocene-Pliocene time, and glaciation during the Pleistocene Epoch.
JACK RATHBONE PHOTO

maroon bells
via co. 82 and
forest route 125

(18-mile round trip from Aspen)

The road to Maroon Bells follows the canyon of Maroon Creek, with high red walls of the Maroon Formation rising above the aspen and spruce of the valley floor. This formation, composed of sand and mud washed off Uncompahgria in Pennsylvanian and Permian time, is often metamorphosed by heat and pressure from surrounding Tertiary igneous intrusions. The metamorphism turns sandstone into harder quartzite and shale into slate; often it causes a graying of the otherwise deep red rocks.

The valley through which you drive is glaciated, the lowest of several moraines being right below the mouth of the canyon, less than a mile from Aspen. A good U-shaped valley profile is visible from there. Hanging valleys and high waterfalls ornament it farther up, and mark the sites where small tributary glaciers entered the main valley glacier. Maroon Lake is dammed by large alluvial fans.

Don't leave without hiking less than two miles to Crater Lake, in the glaciated valley of West Maroon Creek. You'll enjoy seeing one of the most beautiful spots in America, where aspen, spruce, and fir rise against the splendid red-gray cliffs of Maroon Bells, both topping 14,000 feet. Every step reveals new wildflowers by the trail's edge, ground squirrels beg for handouts, and coneys, evasive relatives of the rabbit, watch from a distance or, whistling shrilly, dive into rocky homes. Geologic features abound, from immense rock glaciers visible far up the valley to talus cones building below the avalanche tracks of Maroon Bells.

Between Maroon Creek and Ashcroft in the Elk Mountains, the valley of Conundrum Creek shows the U-shaped cross section that is the signature of glacial erosion. High cirques of small tributary glaciers appear on the right. This valley is accessible only by trail

B.H. BRYANT PHOTO, COURTESY OF USGS

Rock glaciers abound in the Elk Mountains, like this one on Pyramid Peak. Fed with rocks loosened by freezing and thawing, which fall or skid down the mountain slope (background), the mass of broken rock creeps as a unit down the valley.   B.H. BRYANT PHOTO, COURTESY OF USGS

co. 82
aspen — u.s. 24
(44 miles)

At the bridge just southeast of downtown Aspen, Cambrian sandstone can be seen as a ridge running uphill to the right (south). The contact between it and underlying Precambrian granite runs along the east side of the ridge. East of the contact, crossing a moraine, you are in the Precambrian core of the Sawatch Range.

The road goes on up the valley of the Roaring Fork River, which for five miles courses lazily across a wide, flat expanse of glacial outwash. Other moraines close the upper end of the valley, and poorly sorted gravel of old lateral moraines is plastered along the steep slopes near the road. The glacier that occupied this great trough must in places have been thousands of feet thick; at times it rose as high as the angular shoulder far up on the valley wall. Old lateral moraines, as well as rock from the over-steepened canyon walls, have tumbled down toward the valley floor, partly obscuring its U-shaped profile.

Leaving the open valley, the highway climbs the flank of Smuggler Mountain above a series of moraines separated by level meadows (as at Mile 45), and then clings fly-like to a roadway blasted in the white granite of the canyon wall. Far below, the stream becomes a wild tumbling cascade, leaping down a boulder-clogged gorge. In places the canyon walls are smoothed and polished by glaciers. When a glacier occupied this valley, the narrow steep part of the gorge must have been the site of a magnificent icefall, with tumbled blocks of glacial ice separated by deep crevasses.

Close to timberline near the ghost town of Lincoln Gulch, you will see many more glacial features. The cirques behind Lincoln Gulch were carved out of igneous rocks of three peaks of New York Mountain. The last terminal moraines of diminishing glaciers are right in the cirques themselves, rocky lips that emphasize the basin-like nature of the cirques. Hummocky **ground moraines** surface less rugged parts of the upper valley of Roaring Fork River, glacial debris that sank to the valley floor as the last of the glaciers melted away.

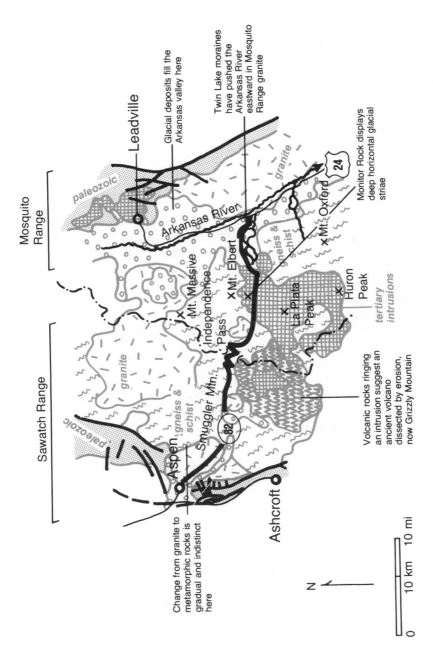

Mosquito Range

Leadville

Glacial deposits fill the Arkansas valley here

Twin Lake moraines have pushed the Arkansas River eastward in Mosquito Range granite

paleozoic

Arkansas River

granite

24

Monitor Rock displays deep horizontal glacial striae

×Mt. Oxford

×Mt. Elbert

gneiss & schist

× Mt. Massive

Independence Pass

×

× ×

La Plata Peak

Huron Peak ×

Sawatch Range

granite

tertiary intrusions

Smuggler Mtn.

82

Aspen *gneiss & schist*

paleozoic

Ashcroft

Volcanic rocks ringing an intrusion suggest an ancient volcano dissected by erosion, now Grizzly Mountain

Change from granite to metamorphic rocks is gradual and indistinct here

N

0   10 km   10 mi

**Colorado 82
Aspen to U.S. 24**

Ancient metamorphic rocks exposed here in the heart of the Sawatch Range are patterned with fascinating wavy bands that tell us the rock must have been nearly molten, perhaps thick and porridge-like, when it was crumpled into mountain roots 1750 million years ago. Pressures that could so squeeze solid rock are attained only deep below the surface of the earth's crust. The granite that intrudes it is associated with a younger set of Precambrian mountains. Once scoured clean by glaciers, both the contorted gneiss and the granite are often covered with fallen rock or thin mountain soils.

The last stretch of road before reaching the pass is a mere ledge cut into tightly folded and fractured and decomposing brown metamorphic rocks, not the most stable of roadbeds. Like many other Colorado passes, Independence Pass seems to have developed where the rock is weakened by fractures, many of which probably date back to Precambrian time. Be sure to look back (if you are not the driver) down the great U-shaped valley of the Roaring Fork. Try to visualize the creeping glaciers that cut these cirques and rounded these slopes, that converge near Lincoln Gulch and slowly but forcefully, over hundreds of thousands of years, shaped the valley all the way to Aspen.

Independence Pass, at 12,095 feet one of the highest highway passes in the United States, is on the Continental Divide. Drainage to the west reaches the Pacific via the Roaring Fork and Colorado Rivers. To the east, Lake Creek (one of the dozen or more Lake Creeks in Colorado) flows into the Arkansas, its waters ultimately reaching the Mississippi and the Gulf of Mexico.

The crest of the Sawatch Range is strongly shaped by glaciers too, and many glacial and alpine features can be observed at Independence Pass. Look for cirques, rolling uplands smoothed by ice, and frost heaving of both rocks and soil on open ground.

Mt. Elbert's summit, carved out of Precambrian gneiss and at an elevation of 14,431 feet the highest spot in Colorado, is just visible due east of Independence Pass, with the round granite summit of Mt. Massive to the left of it. Grizzly Mountain to the south is the core of an eroded Oligocene volcano.

La Plata Peak, to the southeast, has a Precambrian granite summit, but its flanks are eroded from a Laramide batholith composed of porphyry, a rock characterized by large feldspar crystals in a matrix of smaller crystals. This intrusion forms both La Plata and Huron Peaks, two more of Colorado's 53 Fourteeners. Mt. Oxford the north-

213

ernmost of the Collegiate peaks (and also a Fourteener), is made of Precambrian metamorphic rock. It shows just to the left of La Plata Peak. The Sawatch Range claims a total of fifteen Fourteeners and is the highest range in Colorado.

Beyond Independence Pass the highway drops rapidly into the U-shaped glacial valley of Lake Creek. At Mile 62 banded metamorphic rocks and light gray granite are well exposed. Mt. Elbert's massive shoulders rise to the north.

At Mile 73-74, Monitor Rock juts out into Lake Creek Valley from the Elbert Massif. It partly blocked the glaciers that once flowed down the valley, and still bears the scars: deep glacial striae easily visible as parallel horizontal grooves above the highway. Recall that glaciers scour not with ice, but with the rasping action of rocks and boulders frozen in the moving ice.

From Mile 80 the highway looks down on a broadened valley floored with glacial outwash carried by Lake Creek, to Twin Lakes, at the creek's lower end. Lying at the very edge of the valley of the Arkansas River, these lakes are dammed by terminal and recessional moraines of the Lake Creek glacier, with recent modification by man.

Even during the maximum advance of Ice-Age mountain glaciers, the valley of the Arkansas River was not glaciated below Leadville, although it was obstructed from time to time by ice tongues from side valleys. It is floored, however, with terraces of stream-laid glacial gravel, witness to the immense amount of material scoured from these mountains by the glaciers (see U.S. 24 BUENA VISTA — DOWD). Lake Creek Glacier pushed the Arkansas River eastward, forcing it to cut a deep canyon through the granite that forms the lower slopes of the Mosquito Range, now just across the valley from you.

Both the Sawatch and the Mosquito Ranges are parts of the same faulted anticline, with sedimentary rocks dipping west off the Sawatch Range near Aspen, and east off the other side of the Mosquito Range near Fairplay (see U.S. 285 FAIRPLAY — PONCHA SPRINGS). After the uplift was established in Laramide time, its core was split by large faults, now hidden by the valley sediments, and the central unit was dropped down as the northern tip of the Rio Grande Rift, a wedge-shaped slice of the earth's crust that extends far south into New Mexico.

co. 115
colorado springs — u.s. 50
near canon city
(45 miles)

This route, a cut-off to Canon City, skirts the southern end of the Front Range south of Cheyenne Mountain and Pikes Peak, traveling in and out of the band of upturned sedimentary rocks that edge the faulted Front Range anticline. Cheyenne Mountain is to the west as Colorado 115 branches off I-25. To the southeast the Pierre Shale, a greenish gray Cretaceous marine shale not often well exposed, extends across the valley of Fountain Creek. At the foot of Cheyenne Mountain Paleozoic and older Mesozoic rocks have been faulted out, and the Pierre Shale butts against the Precambrian Pikes Peak Granite. The great granite mass of the mountain has been thrust eastward 4000 to 6000 feet over the soft sedimentary rocks.

Near the mountains the Pierre Shale is beveled by a sloping pediment that probably developed in Pleistocene time; it is covered with the remains of a bouldery rockslide that extends like fingers to the southeast. Small conical Tepee Buttes occasionally jut up above the Pierre Shale east of the highway. The resistant rock near their centers contains numerous little fossil clams, and sometimes other fossils.

Toward the south end of Cheyenne Mountain, at Mile 37-36, the Dakota Hogback rises between the highway and the mountain slopes. The fault at the base of Cheyenne Mountain dies out in a series of jogs. The highway begins to cross older and older strata, going down through Cretaceous limestone and hogback-forming Dakota Sandstone (Mile 37) and into Jurassic Morrison Formation, a brightly Easter-egg-colored shale that is easy to recognize. Below the Morrison are some Jurassic layers that contain gypsum, and then the Lyons Sandstone (Mile 36), the smooth salmon-colored sandstone that you may remember from the Garden of the Gods. Shortly beyond Mile 35 the Pennsylvanian Fountain Formation appears, darker red than the Lyons and containing layers of deep red shale.

215

Colorado 115
Colorado Springs to US 50 near Canon City

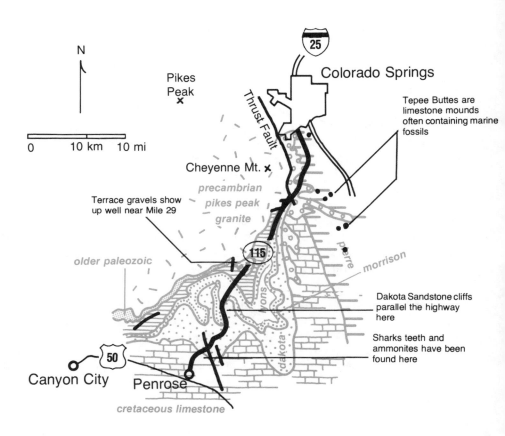

Between here and U.S. 50 near Penrose, the highway crosses back and forth among these rock layers. Close to the mountains they are steeply upturned and the resistant ones, the Lyons, Fountain, and Dakota Formations, cap hogbacks. Farther from the mountains, where dips are gentle, they cap cuestas.

In the distance to the southwest beyond the valley of the Arkansas River rise the Wet Mountains, the east limb of another faulted anticline. Beyond them, blue with distance, is the toothy skyline of the Sangre de Cristo Range, a range with greatest faulting on the west side. Much of the faulting in the Sangres took place in Miocene-Pliocene time, during the regional uplift that raised Colorado and neighboring states to their present mile-high-plus elevations.

Between Miles 25 and 21, a Dakota-capped cuesta parallels the highway. At Mile 20, well around on the south side of Pikes Peak, younger Cretaceous sedimentary rocks begin to appear above the Dakota Sandstone. One is a cuesta-capping limestone on which Co. 115 descends down-slope toward Penrose and U.S. 50. In this limestone sharks teeth, ammonites, and other marine fossils can be found.

High valleys of the San Juan Range, often peppered with mines, reveal the U-shaped cross section characteristic of glacial erosion. Because of over-steepening by glaciers, slopes are now scarred with avalanche and rockslide paths.

JACK RATHBONE PHOTO

v.
fire and ice
— the san juans

The history of volcanism in Colorado really goes back to Precambrian time, but for our purposes let's look only at more recent volcanic episodes, which have left a clearer geologic record. Unfortunately, volcanic rocks don't form the widespread parallel layers that marine sediments do — they thin and thicken, flow down valleys, fill irregularities. And they usually contain no fossils from which their age can be determined. They do, however, bear traces of radioactive materials from which, in the laboratory, their age may be learned.

The Paleozoic Era (Age of Fishes) and most of the Mesozoic Era (Age of Reptiles) passed with virtually no volcanic activity in Colorado. In Cretaceous time some volcanoes must have erupted in the western part of the state, or perhaps in Utah, but nothing is left of them now except occasional thin chemically altered layers of volcanic ash in Cretaceous shale, and rare volcanic pebbles in Cretaceous conglomerate.

During the first part of the Tertiary Period — the time of great mountain-building in the Rockies — there was still little volcanism in Colorado. In the southwestern part of the state a large region was raised into a broad dome 100 miles across, and attacked by erosion that cut clear down into the Precambrian core of the dome.

Then the geologic picture changed abruptly. Beginning about 40 million years ago, volcanoes erupted again and again in the mountain region, in incredible outpourings that lasted, off and on, for 30 million years or more. Ash and lava flows from many volcanoes gradually merged to cover the areas that today are the San Juan Mountains, the West Elk Mountains, and the White River Plateau.

The volcanic region in the San Juans is the most extensive, and in addition the easiest to get to. Here, three classes of volcanic rock are represented:

• Lava flows formed from liquid magma.

• Tuff formed from volcanic ash and pellets of fine volcanic cinder.

• Breccia made of lava fragments — shreds of molten rock thrown from volcanoes, or of crusty lava that repeatedly broke and glued itself together as it crept downslope.

San Juan volcanic rocks have been subdivided into dozens of

formations, but these are difficult to recognize without detailed study. Volcanic activity seems to have come about in three phases, which we'll call the first, second, and third phases, separated by periods of faulting, collapse, and erosion.

The first phase began about 35 million years ago, and lasted five million years or more. In the San Juan region, between 35 and 30 million years ago, lava flows and piles of breccia formed a shield-like volcanic field 100 miles across and perhaps 4000 feet high. Soon after it formed, this shield was blanketed by clouds of volcanic ash that settled thickly over the whole region and extended north to merge with coarse volcanic ash and breccia from the West Elk Mountains. This ash, consolidated now, is called the San Juan Tuff. The volcanoes from which it came have not been identifed for sure, but geologists who have worked in this area believe they were in the northeastern and southern parts of the San Juan area.

A period of erosion followed, and some but not all of the tuff was washed away. What remained was partly covered when more volcanic rocks, coming from volcanoes near the present town of Silverton, built up a tremendous new mound of ash, breccia, and lava flows 40 miles across and 3000 feet high, also considered part of the first phase volcanism.

The ash and lava flows created by these eruptions then slowly settled, probably with the whole terrain sinking into underground magma chambers emptied during eruptions. For a time eruptions ceased and erosion pared the collapsed volcanoes down to mere clustered hills.

Beginning about 29 million years ago, ash and flow eruptions of the second phase began to cover the ruins of the first phase rocks. Again, and indeed time after time, volcanoes erupted violently. Some then collapsed individually to form calderas, circular basins bounded by arcuate faults. Fifteen such calderas have been identified. The volcanic activity soon spread to the central part of the San Juan area, where ash and lava flows from the many volcanoes merged to form a broad volcanic dome. Eight major layers of lava and ash are known. In terms of their composition they represent two kinds of rock: dacite (or rhyodacite) and latite. Dacite is light in color, and like granite contains a lot of silica or quartz, which makes it very stiff so that it forms high, steep volcanoes. Latite, in contrast, is much darker, contains less silica, and is thinner and more fluid when it comes from the ground. It flows quickly over wide areas in nearly horizontal sheets, filling ravines and canyons and leveling the surface. Weathered surfaces of both these rocks appear purplish in the San Juan area.

After the initial second phase eruptions there was another time of collapse and caldera formation. Eruptions from scattered small volcanoes from time to time spread sheets of lava or mounds of breccia on the landscape. Then second phase volcanic activity lessened and, about 26.5 million years ago, ceased. The San Juan region was tilted to the east and to some extent faulted and folded, and with the rest of the region was lifted 5000 feet or more to its present elevation. At about the same time, perhaps 25 million years ago, there came floods of basalt, a dull black, very fluid volcanic rock, that spread out across the San Juan area. It covered most of the eastern part of the San Juans with a basalt veneer, and extended into the San Luis Valley area and 70 miles south into New Mexico as a wide lava plain. These were the third phase eruptions.

The basalt composition of these third phase volcanic rocks suggests that they may have originated much farther below the surface than the earlier volcanic flows, below the silica-rich rocks that are continental in nature, in underlying dark iron- and magnesium-rich rocks similar to those that form the ocean floors. Some of the basalt seems to have risen along the deep faults of the Rio Grande Rift, of which the San Luis Valley in Colorado is a part. This valley developed in Miocene and Pliocene time when the rest of Colorado and parts of adjacent states were being lifted to their present elevations.

Not all the magmas of Tertiary time reached the surface. Around and among the volcanic outpourings of the San Juans are many intrusive bodies: stocks, laccoliths, sills, and dikes. Some of the intrusions are several miles across, and some have lifted overlying sedimentary layers as much as 4000 feet. One of the largest is Sleeping Ute Mountain southwest of Cortez.

The rocks that form many of the stocks and laccoliths, as well as many of the sills and dikes of the area, are fine-grained igneous intrusive rocks that look quite like their extrusive counterparts, the volcanic rocks. Others are porphyries, which have large crystals of some minerals scattered through a finer matrix. Since most of them are in areas where volcanic rocks and overlying sediments have now been removed by erosion, it is difficult to interpret their ages, but they seem to be of about the same age as the second phase volcanic rocks. At the time of their intrusion, the volcanic dome of the San Juans may have extended over them and far to the west, and some of the stocks and dikes may very well have been the conduits through which volcanic magma reached the surface.

Commonly the intrusions baked and hardened and reddened surrounding rock until it resembles fired pottery. Sometimes slow-to-crystallize minerals distilled out of the intrusions, passing into surrounding rocks and enriching them with valuable minerals. Most of the mining towns of the western San Juan Mountains are located near intrusions, and some of the most productive mines are right at the contacts between intrusions and the rocks that surround them.

Though a few more volcanoes erupted in the San Luis Valley during Pleistocene time, the main events in the San Juans involved erosion. Ice-Age glaciers sharpened mountain peaks and carved deep U-shaped, cliff-walled valleys. Streams cut canyons and ravines. Landslides shaped new slopes, dammed streams, and created lakes, giving us the varied San Juan scenery of today.

Above the Gunnison River, West Elk Breccia rests on Mesozoic sedimentary rocks. The craggy cliffs are about 600 feet high.

J.C. OLSON PHOTO, COURTESY OF USGS

u.s. 50

gunnison — montrose

(66 miles)

The most striking geologic features near Gunnison are the Palisades across the Gunnison River northwest of town. These castellated chocolate-colored cliffs, shaped by rain and river, are carved in relatively soft breccia made up of volcanic pumice and ash enclosing boulders and cobbles of harder volcanic rock. The breccia originated more than 25 million years ago in volcanic centers in the West Elk Mountains, which can be glimpsed to the north between Miles 156 and 157.

Southwest of town, rugged hills of Precambrian granite and gneiss appear south of the river, and from Mile 150 the highway is in this Precambrian rock, gray, hard, crystalline, and garlanded with dark and light veins. The country is beginning to open up now, for it is well west of the main Rocky Mountain ranges and will soon spread out into plateau country.

US 50
Gunnison to Montrose

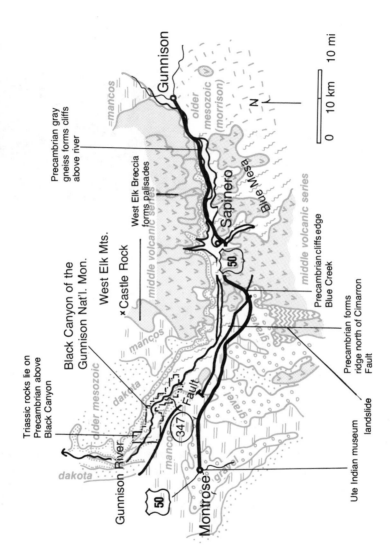

Precambrian gray gneiss forms cliffs above river

West Elk Breccia forms palisades

Black Canyon of the Gunnison Nat'l. Mon.

West Elk Mts.

× Castle Rock

Triassic rocks lie on Precambrian above Black Canyon

Gunnison River

dakota

older mesozoic

mancos

Fault

347

50

grave

Montrose

50

Ute Indian museum landslide

Precambrian forms ridge north of Cimarron Fault

Precambrian cliffs edge Blue Creek

Sapinero

Blue Mesa

middle volcanic series

Gunnison

mancos

older mesozoic (morrison)

N

0   10 km   10 mi

middle volcanic series

Castle-like outcrops of West Elk Breccia contain volcanic fragments imbedded in coarse volcanic tuff. Wind and rain shape these spires and cones, which edge the West Elk Mountains for many miles. W.R. HANSEN PHOTO, COURTESY OF USGS

For several miles the highway runs along the north shore of Blue Mesa Reservoir, at first in the Precambrian rock and then below slopes and cliffs of volcanic breccia similar to that in Gunnison's Palisades. Practically the entire skyline is rimmed with brown volcanic rocks that are derived from either the West Elks or the San Juans. Below, along the highway west of Mile 141, volcanic breccia is underlain by red and green shale of the Morrison Formation. This rock appears frequently in roadcuts, and is easily recognized by its Easter-egg colors. Since it erodes and washes away easily it often helps to undermine the breccia, so it is indirectly responsible for the palisades and cliffs above the highway.

All the sedimentary rock layers between here and Montrose are Mesozoic. Paleozoic rocks, though present here at one time, were stripped off by erosion 200-300 million years ago when this area was part of the Uncompahgria highland of the Ancestral Rockies, an island towering several thousand feet above the surrounding sea. Uncompahgria was gradually eroded, but retained hilly irregularities at least until the middle of the Mesozoic Era. In various places we find it capped by either Triassic or Jurassic formations.

Two other kinds of volcanic rock are visible here, tuff and welded tuff. The tuff consists of fragments of ash or pumice, and forms light-colored hills to the south, as well as upper parts of the cliffs to the north. It is of about the same chemical make-up as granite but with microscopically small crystals.

225

Welded tuff forms during explosive eruptions when extremely frothy volcanic liquids are shattered into tiny glass and pumice fragments by their own rapidly expanding gases. In such eruptions a white-hot cloud of glass fragments and gas bursts from the volcano and shoots, sometimes with terrifying rapidity, down the slopes. (It was this kind of cloud that destroyed Martinique in 1902 and helped to demolish Pompeii in 78 A.D.) Settling, it is still hot enough to weld itself together. Thick welded tuff layers occur on top of some of the West Elk Breccia, and thinner layers are sometimes responsible for funny "hats" and "collars" on pinnacles of breccia.

Massive tuff deposits form whole mountainsides in the San Juans. In places they once were 3000 feet thick, and they must originally have covered an area 100 miles in diameter. Most of the thick tuff layers have now been traced to caldera sources in various parts of the San Juans.

West of Blue Mesa Dam, Precambrian Black Canyon Gneiss abruptly becomes the dominant rock in and near the river. Black Canyon really starts just at the dam, where its walls are about 200 feet high. They become more than 2000 feet high 20 miles farther west. The Black Canyon Gneiss on the nearly vertical canyon walls is a highly variable rock with black mica or hornblende or chlorite giving it its dark appearance, and it is so extensively injected with granite veins that it is sometimes called an injection gneiss.

West of the dam the highway climbs away from the river through Mesozoic sedimentary rocks, to cross a broad and bumpy landslide area with a hilly surface and scattered big blocks of tuff. Such hummocky terrain is characteristic of landslides, and you will soon develop a knack for recognizing it. Slides along the valley here are caused by downslope sliding of the weak volcanic breccia along slippery tuff layers. The slides probably were set in motion by heavy rains, when water absorbed by the porous breccia increased its weight and that absorbed by the tuff layers sort of greased the skids. Highway bumpiness shows the slides still move occasionally. On the north wall of the valley Jurassic and Cretaceous rocks lying on the Black Canyon Gneiss are topped by more cliff-forming tuff, and you can see many landslides there too.

Beyond Mile 124 the highway crosses the erosion-beveled top of Precambrian rocks and drops into Blue Creek Canyon, where these rocks are well exposed. Again they are laced with large and small dikes, some of which contain oversized intergrown crystals of pink quartz and feldspar.

Beyond Blue Creek more hummocky landslides cover the Cimarron Fault, a major fault that has been traced many miles northwest and southeast of here. The position of the fault can be seen from Mile 121: Precambrian rocks north of it form a high ridge that stays in sight for about 10 miles. Cretaceous Mancos Shale slams right into these uplifted Precambrian rocks, as you can see from Mile 114.

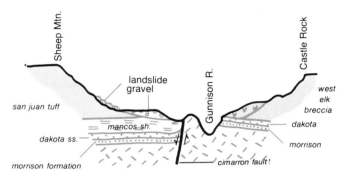

Section across U.S. 50 near Cimarron fault between Gunnison and Montrose.

Notice how bumpy the highway is on Mancos Shale. High clay content of the dark gray shale causes it to swell when wet, a condition that neither pavement nor plants can readily adapt to. The Mancos forms yellow and gray badlands all along the edges of the Uncompahgre Valley, but where it flattens out the shale makes fertile soil.

The rocky summit of Cimarron Ridge is composed of Precambrian rock – Black Canyon Gneiss – while the smoother slopes below are Cretaceous Mancos Shale. The Precambrian of the ridge is lifted along the Cimarron Fault, apparent as a well defined line sloping across the picture from left to right. W.R. HANSEN PHOTO COURTESY OF USGS

227

Near the Black Canyon junction, the faulted ridge of Precambrian rock is veneered with Jurassic shale and sandstone and Cretaceous Dakota Sandstone. These sedimentary rocks slant up eastward, bending over the Precambrian fault block. The relationship can be seen well along the entrance road to Black Canyon of the Gunnison National Monument (itinerary in Chapter VI).

The northern San Juans are now in view to the south, with the San Juan Tuff forming somber cliffs in the middle distance. The Uncompahgre Plateau to the west is a part of ancient Uncompahgria, and was pushed up a second time probably at the time of Miocene-Pliocene regional uplift. Grand Mesa, the lava-capped mesa on the distant skyline north of Montrose, has a similar history, for Uncompahgria was considerably more extensive eastward and southward than the plateau for which it is named.

u.s. 160

alamosa — pagosa springs

(89 miles)

West of Alamosa, U.S. 160 crosses the west edge of the San Luis Valley. For a discussion of this intermontane depression and the great rift valley of which it is a part, see U.S. 285 PONCHA SPRINGS — NEW MEXICO LINE. West of Monte Vista the highway crosses the immense alluvial apron of the Rio Grande, and then near Del Norte plunges into the southeastern San Juan Mountains, a range formed of tier on tier of lava flows and accumulations of volcanic ash.

The west edge of the San Luis Valley is bordered by faults blanketed by thousands of feet of volcanic rock, gravel, and sand. Tongues of volcanic rock reach out into the valley near Del Norte. Volcanic necks, conduits of former volcanoes, jut up sharply around the town, and two large Tertiary intrusive masses surrounded by radiating dikes are a short distance north, all testifying to the magnitude of igneous activity here.

The highway follows the Rio Grande, one of the west's major rivers, as far as South Fork, and then goes up the South Fork to Wolf Creek Pass. Headwaters of the Rio Grande are high above timberline along the Continental Divide well north of U.S. 160. Along this route you

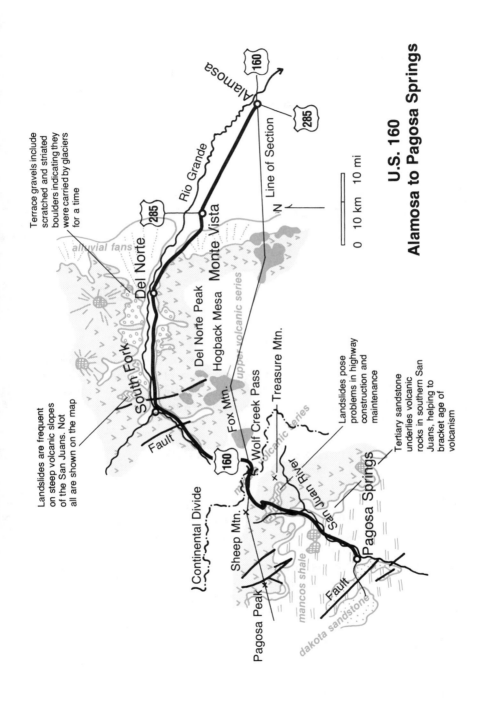

Terrace gravels include scratched and striated boulders indicating they were carried by glaciers for a time

Landslides are frequent on steep volcanic slopes of the San Juans. Not all are shown on the map

Landslides pose problems in highway construction and maintenance

Tertiary sandstone underlies volcanic rocks in southern San Juans, helping to bracket age of volcanism

Alamosa

Rio Grande

Line of Section

N

0 10 km 10 mi

alluvial fans

Del Norte

South Fork

Monte Vista

Del Norte Peak

Hogback Mesa

upper volcanic series

Fault

Fox Mtn.

Treasure Mtn.

Wolf Creek Pass

lower volcanic series

San Juan River

Continental Divide

Sheep Mtn.

Pagosa Peak

mancos shale

dakota sandstone

Fault

Pagosa Springs

U.S. 160
Alamosa to Pagosa Springs

will be rising through many layers of gray volcanic rock dipping gently toward the San Luis Valley.

In this part of the San Juans, third phase basalt lava flows appear on Green Ridge west of Alamosa, on Del Norte Peak and Hogback Mesa southwest of Del Norte, and on Fox Mountain near Wolf Creek Pass. (Del Norte Peak and Fox Mountain are visible from Mile 182-181.) All the rest of the gray or reddish or whitish rocks from Del Norte to Pagosa Springs are lava flows and ash beds of the second phase.

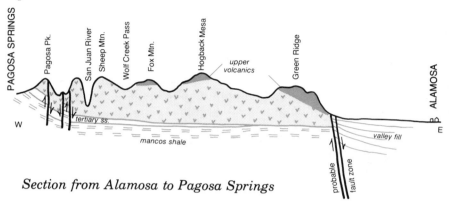

Section from Alamosa to Pagosa Springs

Rocks of the second phase apparently erupted from a number of centers scattered throughout the San Juans, and the series as a whole is made of alternate layers of lava flows and ash beds and breccia interrupted by erosional unconformities. Individual flows and ash deposits may be several hundred feet thick.

As the highway climbs toward Wolf Creek Pass, notice that heavy vegetation covers most rocks. Rich soils derived from volcanic rock combine with fairly heavy rainfall and deep winter snows to nourish a luxurious forest. Frequently, the only visible rocks are the dark lava flows that cap the hills. However, occasional bands of yellow, white, and light gray tuff sometimes show through the trees.

Watch for evidence of glaciation, most apparent in long side valleys. The tunnel at Mile 168 protects the highway from avalanches, which are notoriously severe here where slopes were oversteepened by glaciation.

The area just south of Wolf Creek Pass, around Treasure Mountain, was one of the centers of eruption of second phase volcanoes. It probably was once the site of a tremendous volcano perhaps 65 miles across. White rocks near the pass are ash beds.

230

Descending from the pass, watch along the road for frequent signs of landslides — hummocky surfaces lying below steep scarred slopes or cliffs. Slides have plagued highway construction and maintenance crews here for years, as well as travelers, who have often had to wait while repairs are made. Deep highway cuts in these unstable volcanic rocks almost inevitably start new slides or activate old ones.

The highway zigzags steeply down the slopes of Sheep Mountain on the west side of Wolf Creek Pass. Landslides along this and adjacent mountains, activated by heavy rains, have exposed more volcanic rocks of the second phase. At the viewpoint at Mile 161 you can see some of these rocks up close. They include a coarse fragment-filled volcanic breccia eroded into weird dark pinnacles, and along the edge of the parking lot some shale-like light gray volcanic ash stained with iron minerals. From the viewpoint, bumpy piles of landslide rubble are visible at the base of the mountain face. The San Juan River, coming from the north, winds across a flat valley floor surfaced with outwash from Ice-Age glaciers. The highway soon drops to this level, crossing sedimentary rocks — white Tertiary sandstone (Mile 153) and gray Cretaceous shale, both of which appear in roadcuts.

Near Mile 150 a dike radiating from a little intrusion north of the highway can be seen in the roadcut, offering a close look at the rock of which many San Juan intrusions are made. Clusters of small feldspar and quartz crystals are scattered through the otherwise very fine-grained gray rock. The green mineral that coats some quartz crystals is chlorite. Keen eyes may detect minute pyrite crystals looking like tiny dots of gold.

Just west of Mile 148, a sandy ledge appears in the middle of the dark gray Cretaceous shale where it is exposed across the river. This sandy zone thickens westward to become part of the Mesaverde Group, a group of formations that are several hundred feet thick at Mesa Verde National Park.

Pagosa Springs lies near the west edge of a shale-floored valley, surrounded by hills and cuestas of tilted Dakota and Mesaverde sandstones — both Cretaceous. It gets its name from a cluster of hot springs just south of the San Juan River across the bridge and beyond a group of buildings. Some of the hot water is piped to swimming pools, spas, and fountains, and some is used to heat the nearby motel, but plenty remains in a large morning-glory-shaped pool and several small seeps. In this area, characterized by volcanism of Tertiary and even Quaternary age, the temperature of the earth's crust increases rapidly downward. Groundwater working its way through the rock becomes heated and rises rapidly without cooling off.

The water in these springs contains many minerals in solution, especially silica. As it cools, the chemicals precipitate and accumulate around the springs in the form of **siliceous sinter**. These springs have gradually built a shelf of sinter wide enough to deflect the course of the San Juan River so that it swings westward around them. The rapidity with which the sinter is deposited is evident on the man-made fountains near the motel and across the river, where several feet of sinter have formed in just a few decades.

u.s. 550

montrose — silverton

(60 miles)

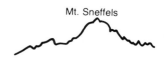

Mt. Sneffels

Leaving Montrose, U.S. 550 follows the Uncompahgre River, crossing gravel terraces made of stream-deposited Ice-Age glacial debris. Visible at valley edges for the first few miles are barren gray and yellow hills of Mancos Shale. Laid down in flat layers in a Cretaceous sea, the shale is now tilted up southward toward the dome that underlies the San Juan Mountains. Its high clay content and its tendency to swell when it gets wet discourage all but a few extra hardy plants.

Almost all of the San Juan skyline ahead is composed of first phase volcanic rocks. However, Mt. Sneffels, the most prominent peak to the south, is a small Tertiary intrusion, and Potosi Peak just east of it is topped with second phase lava flows. Wetterhorn, the horn-shaped peak farther east, is another intrusive peak.

At Mile 116 the Dakota Sandstone comes to the surface and forms ridges on both sides of the river. Below it, especially farther south near Mile 113, you will see green and purple shale of the Jurassic Morrison Formation. Still farther, near Ridgway, these rocks vanish as the highway crosses a fault, and then the Mancos-Dakota-Morrison sequence is repeated.

Older Jurassic rocks surface as you continue south, among them the distinctive light-colored "slick rim" of the Entrada Sandstone, an ancient dune deposit that now forms a unit easily identified all over southwest Colorado and adjacent parts of Utah.

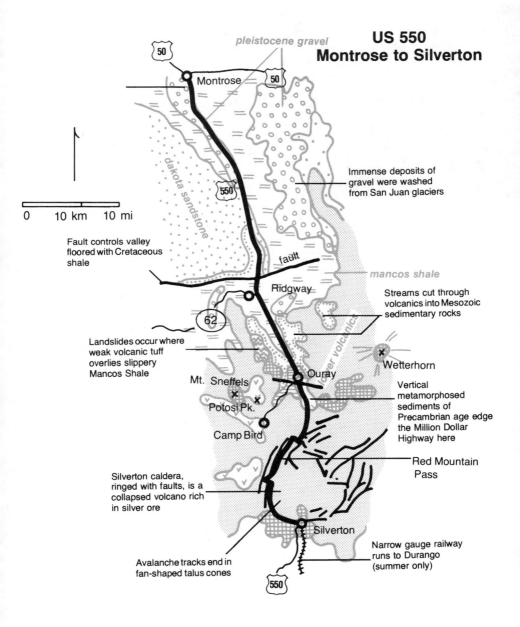

pleistocene gravel

US 550
Montrose to Silverton

50

Montrose 50

550

Immense deposits of
gravel were washed
from San Juan glaciers

dakota sandstone

0   10 km   10 mi

Fault controls valley
floored with Cretaceous
shale

fault

mancos shale

Ridgway

Streams cut through
volcanics into Mesozoic
sedimentary rocks

62

lower volcanics

Landslides occur where
weak volcanic tuff
overlies slippery
Mancos Shale

Wetterhorn

Ouray

Mt. Sneffels

Vertical
metamorphosed
sediments of
Precambrian age edge
the Million Dollar
Highway here

Potosi Pk.

Camp Bird

Red Mountain
Pass

Silverton caldera,
ringed with faults, is a
collapsed volcano rich
in silver ore

Silverton

Avalanche tracks end in
fan-shaped talus cones

Narrow gauge railway
runs to Durango
(summer only)

550

233

Mt. Sneffels is an eroded mass of intrusive igneous rock cutting up through surrounding layers of volcanic rock. Slanting talus slopes are fed from the irregular summit cliffs, and in turn nourish a rock glacier in the cirque below the peak. W. CROSS PHOTO, COURTESY OF USGS

Triassic and Permian rocks appear south of Ridgway also, dark red sandstone and shale which form slopes and ledges below the "slick rim." Difficult to date, for they contain few fossils, these rocks form layer-cake cliffs near Ouray. Uppermost Pennsylvanian rocks are red also, so the whole "'redbed" sequence spans about 120 million years. During that time the ancient highland of Uncompahgria yielded little by little to forces of erosion and decay. The many streams draining it deposited layers of red mud and sand on deltas and floodplains perhaps similar to those of the Mississippi or Nile today.

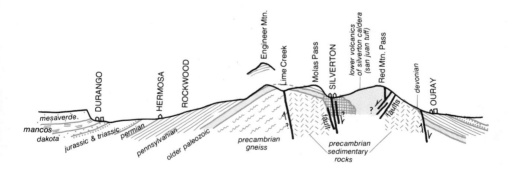

Section across San Juan Mountains parallel to U.S. 550 Ouray to Durango.

Ouray's municipal swimming pool is supplied with naturally warm water by hot springs just north of the town. Ouray was settled in 1875 when gold and silver were found near Mt. Sneffels. It may at the time have seemed a logical and strikingly beautiful place for a town, but avalanches and landslides do not make it the safest place to live. The district has produced over $125,000,000 in gold, silver, lead, copper, and zinc. The richest deposits lie near Camp Bird Mine about six miles southwest of Ouray, and at Red Mountain 12 miles south.

Below the Triassic, Permian, and Pennsylvanian redbeds in the Ouray cliffs, and also rising southward onto the San Juan dome, are Lower Paleozoic marine sedimentary rocks. The most distinctive of these is massive gray Leadville Limestone. Below it are older, browner limestone layers called the Ouray Formation. Ouray's Box Canyon Falls occur where Canyon Creek, approaching from the southwest in a hanging valley, has dissolved a deep cleft in fault-weakened limestone, with walls that overhang the falls by nearly 100 feet. Formed of the mineral calcite, limestone is surprisingly soluble in the slightly acid water that has percolated through volcanic rock and forest soil.

Devonian sandstone layers lie almost horizontally across beveled vertical Precambrian metamorphic rocks near Box Canyon Falls. Distant crags are eroded in San Juan Tuff.
JACK RATHBONE PHOTO

235

Just above the falls, south of the fault, Devonian sandstone lies almost horizontally across Precambrian rocks, on the erosion surface formed during the long interval between Precambrian and Devonian time. Notice that here the Precambrian rocks don't look like the gneiss and granite that you see elsewhere in Colorado. They are clearly sedimentary rocks — conglomerate, sandstone, and shale — slightly metamorphosed into quartzite and slate and schist. They are of younger Precambrian age than the gneiss and granite seen elsewhere in Colorado, but their vertical position still dates from Precambrian time.

The oldest volcanic deposits here are the San Juan Tuff, a thick widespread layer of volcanic ash and fine breccia. The large amphitheatre above Ouray is formed almost entirely of this first phase rock.

South of Ouray U.S. 550 becomes the historic "Million Dollar Highway," built back when $1,000,000 built 50 miles of mountain road. The highway was blasted through light Precambrian quartzite and dark slate and schist, steeply dipping and in many places rounded by Ice-Age glaciers. Beyond Mile 90, gray volcanic rocks overlie an irregular surface on Precambrian rocks; all Paleozoic rocks were eroded off this part of the San Juan dome before the volcanic rocks were deposited.

Near Ironton the highway emerges onto a flat lake floor filled with sediments laid down when the valley was dammed by a landslide. The many mines along the highway near Red Mountain Pass worked small pipe-like ore bodies very rich in silver-copper and silver-lead ores. Underground this area is honeycombed with mine tunnels! Gaudy colors of surface rocks result from oxidation of iron-bearing minerals, not concentrated enough to be mined, but often used by prospectors and geologists as clues to the whereabouts of other more valuable minerals.

At Red Mountain Pass the route enters a remarkable area, the collapsed cauldron or **caldera** of an ancient volcano. A caldera forms by collapse of the whole summit of a volcano, which sinks along a circular array of faults into partly emptied magma chambers underneath. True to form, the Silverton caldera is ringed with faults. But don't expect to see an obvious caldera like those at Crater Lake or Haleakala. This one was covered over by more volcanic outpourings, and glaciers much later carved up the mountains disregarding the volcanic structure, so the outline of the caldera is not apparent on the ground. Laced with radial fissures and dikes, impregnated with minerals, the Silverton caldera has produced more than $150,000,000 in metals, mainly silver and gold.

Silverton's mining history began in 1870, when gold was discovered on what was then Ute Indian territory. Real mining began in 1874 after a treaty was signed and ratified, but nearly 4,000 claims had already been recorded! Despite the remoteness of the area and the hardships involved in high-altitude mining, Silverton flourished even before the narrow-gauge railway came through from Durango in 1882. In summer now the town revives the atmosphere of the gold camp of the roaring 80s, with costumed residents crowding to the station around noon to watch the train come in. For the narrow-gauge still winds through rugged, lonely Animas Canyon, where mine-dotted crags testify to the lure of silver and gold and the tenacity of fortune hunters. A small museum in Silverton's City Hall displays local minerals, and jeep tours carry visitors to old mines.

Between Red Mountain Pass and Silverton, the highway descends the U-shaped glaciated valley of Mineral Creek, its glacial profile modified by red and yellow talus slides and avalanche chutes. Above the center of the photograph, a large rock glacier creeps slowly down a steep ravine between Grand Turk and Sultan Mountains. L.C. HUFF PHOTO, COURTESY OF USGS

US 550
Silverton to Durango

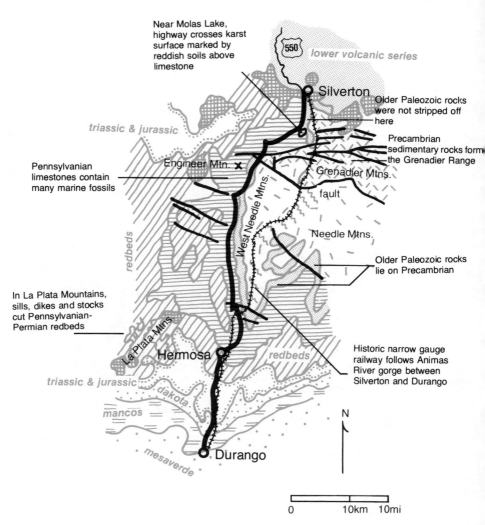

Near Molas Lake, highway crosses karst surface marked by reddish soils above limestone

lower volcanic series

Silverton

Older Paleozoic rocks were not stripped off here

Precambrian sedimentary rocks form the Grenadier Range

triassic & jurassic

Engineer Mtn. **X**

Grenadier Mtns.

Pennsylvanian limestones contain many marine fossils

West Needle Mtns.

fault

Needle Mtns.

redbeds

Older Paleozoic rocks lie on Precambrian

In La Plata Mountains, sills, dikes and stocks cut Pennsylvanian-Permian redbeds

La Plata Mtns.

Hermosa

redbeds

Historic narrow gauge railway follows Animas River gorge between Silverton and Durango

triassic & jurassic

dakota

mancos

N

mesaverde

Durango

0   10km   10mi

238

u.s. 550
silverton — durango
(51 miles)

Leaving Silverton, U.S. 550 passes through about a mile of gray Tertiary intrusive rock, then climbs toward Molas Pass through Cambrian and Devonian sediments into massive red-stained Leadville Limestone. It eventually rises into Pennsylvanian dull dark purple shale full of broken pieces of Mississippian limestone. All these Paleozoic rocks, though they are high on the west side of the San Juan dome, somehow escaped erosion before and during and after the San Juan volcanic extravaganza.

At the top of the Leadville Limestone the surface is distinctly irregular, a **karst** surface (named after a region in Yugoslavia) caused by solution of the limestone at the end of Mississippian time. Karst areas exist today in Yugoslavia, Puerto Rico, and parts of southeastern United States — the ingredients are limestone and a wet, warm climate. They are regions full of solution caves, collapsed sink holes, and extremely irregular and broken land surfaces usually coated with red soil. There is good evidence that such a surface covered much of Colorado just before Pennsylvanian seas advanced across the land. Some of the Pennsylvianian rocks here contain rounded pebbles of hard black chert, a mineral that consists of a mass of extremely small quartz grains. Mississippian fossils in the chert show that the pebbles came from chert layers in the Leadville Limestone.

The slopes above Molas Lake are marked with horizontal limestone beds in Pennsylvanian marine strata. The ledge-forming limestone layers contain abundant fossil shellfish. The Pennsylvanian rocks are cyclic, with repeated sequences of sandstone, shale, and limestone. Cycles typify sediments of this age all over the world. In Illinois, Pennsylvania, and other eastern states, as well as in England and other parts of Europe, the cycles contain thick coal beds formed in ancient swamps. Many attempts have been made to explain the cause of the cycles; one fairly plausible explanation is that they are due to changes in sea level caused by repeated ebb and flow of arctic or antarctic glaciation that lowered sea level by tying up

239

Molas Lake and a number of smaller ponds fill hollows in the old karst surface on top of the Leadville Limestone. The Needle Mountains, visible across the lake, are made of Precambrian granite and gneiss of the center of the San Juan dome. L.C. HUFF PHOTO, COURTESY OF USGS

moisture in masses of ice. Another is that they are due to sea level changes caused by repeated movements along mid-ocean ridges.

The Grenadier Mountains northeast of Molas Lake are made of steeply dipping layers of Precambrian quartzite and slate similar to those near Ouray. Needle Mountains, visible to the southeast across the lake, are Precambrian granite and gneiss.

The flattened tops of some of the peaks near Molas Pass were at one time considered remnants of a widespread, nearly featureless erosional surface which geologists at that time called the San Juan Peneplain. But recent studies show that they are lava and ash flows and not old erosion surfaces at all. Surrounding high rolling country was smoothed and scoured by an immense Pleistocene icecap through which projected a few nunataks — Mt. Sneffels, Engineer Mountain, and other jagged peaks not rounded by ice. There is certainly abundant evidence of glaciation in cirques, U-shaped valleys, smoothed rounded rock surfaces, and occasional moraines. Watch for glacial striae on hard rock surfaces.

At Mile 61-60 the highway crosses a small segment of Precambrian metamorphosed sediments — limestone recrystallized into marble, shale compressed to slate, and sandstone cemented into quartzite. These rocks occur in a narrow fault slice that extends from the west

Cyclic Pennsylvanian marine sediments form the lower slopes of Engineer Mountain. Pennsylvanian and Permian redbeds – sandstone and shale washed from ancient Uncompahgria – form the higher slopes. The cliff on the right side of the mountain is an igneous sill, probably of Tertiary age. JACK RATHBONE PHOTO

end of the Grenadier Range. Crossing another fault, the road cuts into a Tertiary intrusion (Mile 60) with gray rock full of scattered needles of the mineral hornblende. Then more Pennsylvanian cyclic sediments, which include black shale easily mistaken for coal near Coal Creek. A fold in these cyclic layers is visible at Mile 59.

Older Paleozoic sediments appear in a roadcut at Mile 54 — the Mississippian limestone with its broken karst surface, the brownish Devonian limestone below it, and then Cambrian rocks. (Remember — there is no Silurian in most of Colorado. Ordovician is absent here as well.) These sedimentary rocks lie on dark Precambrian gneiss with pink dikes, and they dip southwest here, off the south side of the San Juan dome. In places, the highway runs right on the Mississippian karst surface, recognizable by its irregularity and bright red soils. Pennsylvanian rocks appear in the slopes to the right; Mississippian and older Paleozoic rocks are below the road, to the left.

The West Needle Mountains, visible east of the highway between Miles 60 and 50, are made of Precambrian metamorphic rock, mostly gneiss. The Animas River (out of sight from the road) cuts down through the gneiss into still older metamorphic rocks, to separate the Needle and West Needle ranges. The narrow-gauge railway between Durango and Silverton threads through the Animas River gorge or clings precariously to the canyon walls.

241

U.S. 550 crosses more karst Mississippian surface at Mile 41, on very fossiliferous and often cherty limestone. (This is the chert that sometimes ends up as pebbles in Pennsylvanian conglomerate.) Then it drops down through older Paleozoic rocks, usually concealed by vegetation.

As the highway's grade decreases, it begins to pass gradually back into younger rocks, starting at Mile 36, where it crosses from Cambrian whitish sandstone to brown Devonian limestone in which large calcite crystals twinkle in the sunlight. By Mile 30 you will have crossed Mississippian limestone and the Pennsylvanian marine sandstone-shale-limestone cycles, and come onto Pennsylvanian and Permian redbeds.

All these rocks have been given formation names, many of them derived from places along this road: Ouray Limestone (Devonian), Molas Formation (red karst soils), and Hermosa Formation (Pennsylvanian marine cycles) are examples. Near Hermosa the Hermosa Formation is 2000 feet thick, and includes greenish gray sandstone and conglomerate, thin beds of fossil-bearing gray and black limestone, and some dark gray or reddish shale. Thin beds of gypsum indicate that the area was an isolated basin from which seawater sometimes evaporated.

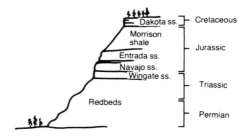

Permian and Mesozoic rocks above Mile 26, U.S. 550 Silverton to Durango.

Somewhere near Mile 28 Permian redbeds give way to Triassic ones, but the level at which this change occurs is difficult to determine since the rocks contain few fossils. You should be able to recognize the Entrada Sandstone "slick rim" and the Easter-egg colors of the Morrison Formation, noted for its dinosaur bones and polished gastroliths or gizzard stones, stones swallowed by the ancient reptiles as an aid to digestion. Above the Morrison is the Cretaceous Dakota Sandstone, widespread and resistant, forming the canyon rim.

Durango is on the edge of the Plateau country of southwestern Colorado. Sedimentary layers flatten out here, forming cuestas, mesas, and plateaus, all dipping gently south into the San Juan Basin.

Durango was for many years Silverton's chief contact with the outside world. The narrow-gauge railway that connected these towns continued southeast to Chama, New Mexico, and then northeast to Antonito and the standard-gauge. Boomtown days are over as far as mineral shipments and mining supplies are concerned; today's boom is in the tourist trade, and diminutive locomotives and cars of the narrow-gauge carry sight-seers and vacationers.

co. 145

telluride — cortez

(72 miles)

In most parts of Colorado, it was elemental or native gold that first lured prospectors and miners. This was true in the western San Juans too, but later prospecting revealed that this area also contained a gold-bearing tellurium compound called **telluride**. The ores for which the town of Telluride was named occur in veins associated with many small intrusions, veins which contain both native gold and compounds of lead, silver, zinc, and gold. Telluride's miners went to great pains to get at these ores. Some veins have been mined vertically for nearly 3000 feet; others have been followed horizontally for close to seven miles.

Most of the original mine portals are now covered by second-growth aspen and fir. The original forest was cut long ago for mine timbers, log cabins, and fuel. Many old tunnels now interconnect and can be reached through the Idarado mine at the head of the valley.

Leaving Telluride the highway passes "Society Turn," where Telluride's "carriage set" rode on Sunday afternoons, and goes up South Fork's typical U-shaped glaciated valley. A cliff-forming 800-foot sill across this valley connects the igneous rock of the Ophir and Mt. Wilson intrusions. Dakota Sandstone above the sill was lifted well above its normal position by the intrusion.

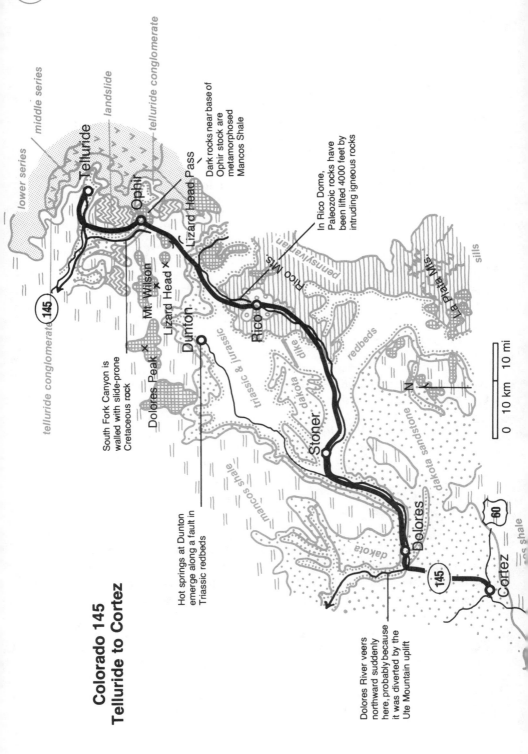

Colorado 145
Telluride to Cortez

lower series

middle series

landslide

telluride conglomerate

telluride conglomerate

Telluride

Ophir

Lizard Head Pass

Dark rocks near base of
Ophir stock are
metamorphosed
Mancos Shale

In Rico Dome,
Paleozoic rocks have
been lifted 4000 feet by
intruding igneous rocks

Rico Mts.

Mt. Wilson ×

Lizard Head ×

Dolores Peak ×

Dunton

South Fork Canyon is
walled with slide-prone
Cretaceous rock

145

Rico

dakota

dike

pennsylvanian

redbeds

La Plata Mts.

sills

triassic & jurassic

Stoner

N

0 10 km 10 mi

Hot springs at Dunton
emerge along a fault in
Triassic redbeds

mancos shale

dakota

Dolores

dakota sandstone

60

mancos shale

145

Cortez

Dolores River veers
northward suddenly
here, probably because
it was diverted by the
Ute Mountain uplift

Near Ophir the most prominent geologic feature is the Ophir stock, a huge mass of granite-like rock that contains big pink and white feldspar crystals and smaller crystals of quartz, hornblende, and mica. Overlying sedimentary and volcanic rocks, shattered and tipped up vertically as a result of the intrusion, are eroded into the spectacular pinnacles of Ophir Needles. The Ophir district once produced gold, silver, lead, zinc, and tungsten; it is almost inactive now.

A small hanging valley cuts through part of the Ophir stock, and glacially polished, grooved, and striated rock can be seen along both sides of the dirt road to Old Ophir. The glaciers that occupied this valley during the Ice Age were unable to carve downward as rapidly as the large South Fork glacier, partly because of their smaller size and partly because of the hardness of the Ophir intrusion, so their valley was left high above the main one when the last glaciers melted.

From Mile 67, Colorado 145 Telluride to Cortez

Large landslides down and to the right of the highway below Mile 64 have destroyed portions of the tracks and trestles of the old narrow-gauge railway from Telluride to Durango. Landslides are also responsible for Trout Lake, for the timber-covered slopes east and northeast of the lake are blocks of San Juan Tuff that slid down from the main ridge line more or less intact, skidding on slippery Cretaceous shale.

At Lizard Head Pass more features of the region come into view. The rugged ridge of Yellow Mountain to the north has slopes of San Juan Tuff colored by oxidation of iron minerals as igneous vapors seeped along veins and fractures. Its jagged skyline is cut in younger lava flows belonging to the second volcanic phase. The volcanic rocks lie on Cretaceous shale that occurs near the pass and forms the smooth green slopes of Sheep Mountain nearby. No third phase volcanics occur in this part of the San Juans.

More smooth shale slopes rise to the large, very irregular intrusion that forms the high peaks of the San Miguel Mountains, northwest of Mile 58. Peaks west of the narrow spine of Lizard Head are parts of this intrusion, though Lizard Head itself, and parts of Mt. Wilson, are eroded remnants of pre-volcanic conglomerate and reddish gray San Juan Tuff.

Much of the high upland here is glaciated, but the sharp, pinnacled silhouettes of many of these peaks suggest that they jutted up through the ice as nunataks. Evidences of glaciation continue almost to Rico. Their lower limit, at about 8000 feet in most parts of Colorado, is above 9000 feet here because of the southern exposure. Terraces of river-deposited gravel washed from the glaciers of course extend much lower.

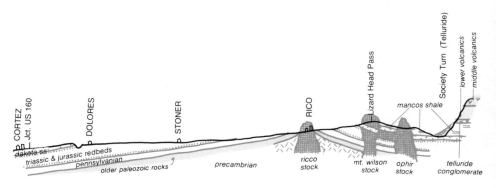

Section along Colorado 145 from Society Turn near Telluride to Cortez

Between Lizard Head Pass and Rico the highway goes through older and older sedimentary rocks that are rising to cross a dome in the vicinity of Rico. In the order in which you will encounter them, they are:

• Dakota Sandstone (Mile 56), the lowest Cretaceous unit in most of Colorado, the familiar tan, resistant, blocky sandstone that often tops cuestas and hogbacks.

• Jurassic sedimentary layers about a thousand feet thick, including floodplain, lake, tidal flat, estuary, and dune deposits. The uppermost unit is the Morrison Formation, recognizable by its Easter-egg colors. The lowest is the Entrada Sandstone, a dune sandstone that forms an unmistakable smooth, light-colored, rounded rock ledge locally called the "Slick Rim" (Mile 54).

• Triassic and Permian red sandstone and shale laid down in conditions that allowed oxidation of iron, giving the rocks a red color. These are probably floodplain, delta, and estuary deposits formed very near sea level. But an important geologic event is recorded near

The Entrada Sandstone erodes into a smooth, often rounded cliff locally called the "slick rim." Here it displays the distinctive diagonal cross bedding that indicates it is a dune deposit. E.B. ECKEL PHOTO, COURTESY OF USGS

their base, for a conglomerate there contains pebbles and cobbles of Precambrian igneous rock, indicating that somewhere nearby the land had lifted so high that erosion could cut into the underlying granite. The conglomerate, then, signals the surge of uplift that created the Ancestral Rocky Mountains late in Pennsylvanian time. Its pebbles almost certainly come from Uncompahgria, the island range of southwestern Colorado.

• Pennsylvanian marine shale and marine limestone containing fossils.

In the rugged Rico Mountains surrounding the town of Rico, an intrusion of molten igneous magma pushed up strongly enough to arch these sediments to form the Rico dome. Total upward doming was at least 4000 feet. Many sills and dikes and irregular bodies of igneous rock occur here over an area about five miles across. Some of them may have been the conduits of a cluster of volcanoes. Near the town of Rico the intrusive rocks are mostly concealed by landslides.

And Rico itself is built on a landslide! It was once a silver-mining town, obtaining silver ore from Pennsylvanian rocks near their contact with Tertiary intrusions. More recently, lead, zinc, and pyrite have been mined here.

South of Rico the rocks dip south off the dome, so that as the highway travels down the Dolores River it goes **up** through the rock sequence from Pennsylvanian to Cretaceous, reversing the sequence given above.

Notice how the Dolores River unexpectedly turns north beyond Dolores. Some geologists think it once continued southwestward off the slope of the San Juan uplift, but was diverted from its original course when intrustions of molten magma pushed up the Ute Mountains southwest of Cortez. The river now flows far northward through tortuous high-walled canyons to join the Colorado River in Utah.

The last leg of this stretch of highway crosses a gently sloping surface of Dakota Sandstone to join U.S. 160 two miles east of Cortez.

co. 149

south fork — blue mesa reservoir

(123 miles)

The community of South Fork is surrounded by hills of welded tuff of the second volcanic phase, well exposed northeast of the highway beyond Mile 4. The welded tuff was deposited as broadly domed layers alternating with loose or unwelded tuff beds. Other domes were smaller and consisted of more irregular, bumpier flows and beds of broken breccia. Because the volcanic domes have been extensively carved up by erosion, they are recognizable now only by careful geologic mapping of the thicknesses of the different units.

The distinctly V-shaped canyon of the Rio Grande between South Fork and Creede was not glaciated. Its shape is governed by huge rockfalls where the canyon walls are undermined by river and road. As you can see, rockfalls pose a major problem for highway maintenance crews here.

Just beyond Wagon Wheel Gap a road to the right leads eventually to Wheeler Geologic Area, a wonderland of pink and white spires, cones, and gnome-like figures eroded in relatively soft second-phase tuff. A four-wheel-drive vehicle is a **must** for this rocky 24-mile trip. Though the variety of form and color is unsurpassed in any other one area, small clusters of similar spires can be seen in other parts of the San Juans.

North of Wagon Wheel Gap, flow structure in the volcanic rocks across the river shows up so well it is difficult to believe the rocks are

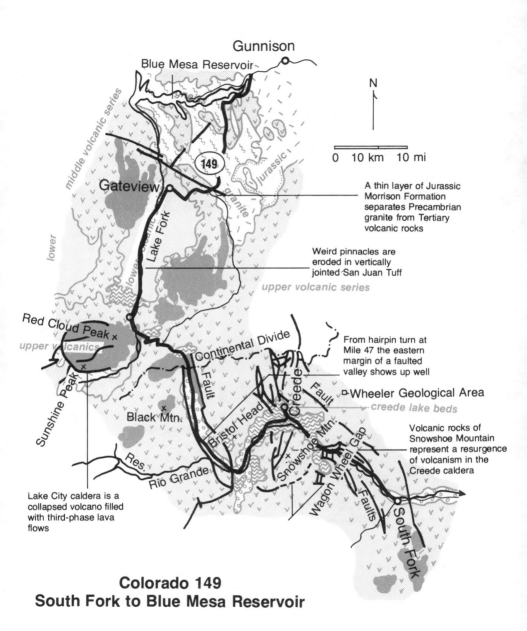

Gunnison

Blue Mesa Reservoir

N

0   10 km   10 mi

middle volcanic series

Jurassic

granite

A thin layer of Jurassic
Morrison Formation
separates Precambrian
granite from Tertiary
volcanic rocks

149

Gateview

Lake Fork

lower

lower volcanic

upper volcanic series

Weird pinnacles are
eroded in vertically
jointed San Juan Tuff

Red Cloud Peak ×

upper volcanics

Continental Divide

Sunshine Peak

×

From hairpin turn at
Mile 47 the eastern
margin of a faulted
valley shows up well

Fault

Creede

Fault

Wheeler Geological Area

creede lake beds

Black Mtn. ×

Bristol Head

+

Snowshoe Mtn.

+

Volcanic rocks of
Snowshoe Mountain
represent a resurgence
of volcanism in the
Creede caldera

Res.

Rio Grande

Wagon Wheel Gap

Faults

South Fork

Lake City caldera is a
collapsed volcano filled
with third-phase lava
flows

**Colorado 149
South Fork to Blue Mesa Reservoir**

At Wagon Wheel Gap, the Rio Grande flows below rugged basalt cliffs that show gently sloping lava flow surfaces and irregular columnar jointing. This photograph was taken by W.H. Jackson in 1874, long before the present highway was built. COURTESY OF USGS

25-30 million years old! Columnar joints caused by cooling and shrinking of the lava make this flow look like an old log stockade.

One of the second-phase volcanic domes, more than 2000 feet high, centered near the town of Creede. This dome apparently blew its top or collapsed, forming a caldera nearly 10 miles across. In a crescent-shaped lake similar in origin to Crater Lake in Oregon, fine yellow and white mud made largely of volcanic ash accumulated. The fine sediment, now turned into yellow and white platy shale, preserved many Oligocene fossil leaves and insects. They are not difficult to find; look for them in the white shale near the highway just west of Creede.

Abundant faults in the Creede area probably are related to the collapse of the Creede volcano. Highly mineralized, they have made the Creede district one of the most productive silver-mining areas in the United States. Discovered in 1889, the district survived the 1893 silver crash and went on to produce gold, lead, and zinc as well. Ores occur in quartz and amethyst veins in shattered volcanic rocks north of the town. Drive up past the fork in the canyon to see the mines, some of them real cliff-hangers. Mill wastes fill the valley south of Creede.

Three or more easily identified terrace levels show up along the Rio Grande west of Creede. Each level, part of a former river floodplain, represents a time of stability in the history of the river, when it was

After the Creede volcano blew up or collapsed, a new volcanic dome developed in the center of its wide caldera. This dome is now visible as Snowshoe Mountain, south of Creede. P.W. LIPMAN PHOTO, COURTESY OF USGS

abundantly supplied with both gravel and water. The uppermost terrace is the oldest (see I-76 FORT MORGAN — DENVER). Higher terrace-like benches visible on slopes in the distance appear to be ancient lake shorelines carved into the surrounding volcanic rocks.

West of Mile 29 the road leaves the last of the lakebeds and enters more second-phase volcanic rocks. They are well exposed on the high cliffs of Bristol Head, visible to the right from Mile 32. At about Mile 30 the road swings northward up a broad valley on which the Rio Grande swings back and forth in lazy meanders. This valley was not shaped by the Rio Grande, which actually enters it from the west. It is a structural valley, a down-dropped block about 20 miles long and one to four miles wide, faulted along both sides, with the center block 2000 feet lower than the bordering highlands. Bristol Head and the cliff-like ridge running north from it are the uplifted eastern side. The highway crosses the eastern fault at about Mile 30, but here and for much of the valley's length both fault zones are hidden by valley gravel. The western side is a broad mesa except where it is cut by the glacier-carved upper Rio Grande Valley, best seen from the viewpoint near mile 48. The central downdropped block is tipped, and tilts up eastward, in part hiding Bristol Head and cliffs north of it as well as another narrower fault-controlled valley. The eastern uplifted margin of the main valley shows up well from the hairpin turn at Mile 47.

At Mile 53, volcanic rocks capping the two ridges to the northeast belong to the third volcanic phase. The highway crosses some rocks of this series at the Continental Divide. They differ markedly from those of the second phase, for they cut across older rocks and form hard caprock over distances of several miles.

Beyond the Continental Divide the highway swings across rolling highlands edged with slumps or earthflows, characterized by hummocky and often aspen-covered terrain. This whole region is under-

251

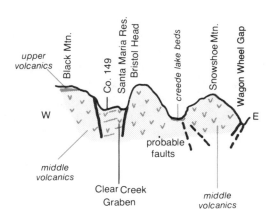

Section from Wagon Wheel Gap to Black Mountain across Clear Creek fault valley

lain by weak tuff and breccia of the first volcanic phase. The most recent of the slumps, Slumgullion Earth Flow, is visible from the overlook between Miles 65 and 66. Trees at drunken angles, and surveys made annually indicate that the flow still moves sporadically. It dams San Cristobal Lake near Lake City, and a lakeside resort (built no doubt without geologic advice) seems to dare it to move faster! Where the highway crosses the earth flow, the bumpy pavement is another indicator of recent movement.

Red Cloud and Sunshine Peaks can be seen now to the west. They occupy a caldera somewhat like that at Creede. In this case the shape of the caldera is well defined by the ring of surrounding peaks. The central peaks and much of the caldera rim belong to the third volcanic phase. However the south and west sides of the caldera are edged with some Precambrian rock that is part of the uplifted core of the San Juan Range.

Uncompahgre Peak, the highest peak in the San Juan Range, is also visible from Slumgullion overlook. It is topped with second-phase volcanic rocks. Sharp-pointed Wetterhorn, a small intrusion surrounded by radial dikes, probably was once the conduit of a large volcano.

Lake City mines produced both gold and silver. The town now welcomes summer tourists, and jeep tours visit mining sites and old ghost towns.

Below Lake City the road is bordered with bluffs of tuff and breccia of the first volcanic phase. Seen up close the rock is gray or purplish gray and coarser than it appears at a distance, with numerous small fragments imbedded in it. Most of it came from volcanoes as ash and cinder, and parts of it, particularly where horizontal banding is strong, may have been water-laid.

The area around Lake City and Gateview is one of the few places where first-, second-, and third-phase volcanic rocks appear together. Third-phase flows top the mesas northwest of Mile 89. Second-phase layers form slopes and ledges below. And tuffs of the first phase edge the Lake Fork of the Gunnison River. In many places the early tuffs shrank as they cooled and fractured into vertical columns. Where weathering is controlled by the pattern of the vertical joints, fantastic pointed cones have developed. Though often called "tent rocks," they look more like skinny tepees.

At Mile 103-104 the highway crosses the base of the volcanic rocks. Breccia and tuff rest on thin layers of colorful Jurassic shale that in turn lies on Precambrian granite and gneiss. There are no Paleozoic sedimentary rocks here, for this area was part of Uncompahgria and if they existed here at one time they were eroded off. Prospect pits for uranium dot the shale outcrops. Precambrian rocks, dark hornblende gneiss garlanded with pink dikes and irregular masses of granite, belong to the Black Canyon Gneiss. They are particularly well exposed between Miles 106 and 109.

The West Elk Mountains, now visible on the northern skyline, are another volcanic center, with rocks similar to those of the first phase in the San Juans. Some of the tallest peaks are volcano conduits surrounded by radiating dikes. From them lava flows slope toward the south, interlayering with the San Juan Tuff. The two volcanic fields are separated by the Gunnison River, which during much of Tertiary time must have been pushed alternately south and north by eruptions in one or the other of the two volcanic fields (see BLACK CANYON of the GUNNISON NATIONAL MONUMENT).

Colorado's western plateau area, about a fifth of the state, is sculptured into deep canyons and ravines that separate isolated mesas, buttes, and plateaus. The shape of the land is governed by hardness and softness of horizontal rock layers. Mesas and plateaus are capped with resistant sandstone layers, or in some instances with lava flows; their sloping sides are often eroded in soft non-resistant shale.
JACK RATHBONE PHOTO

vi.
land of lonesome beauty — the plateau country

West of the Rocky Mountain ranges is a region of flat-lying sedimentary rock that is part of the Colorado Plateau Province, a vast area that extends well into Utah, Arizona, and New Mexico. Somehow this region escaped the wave of unrest that swept from the west across western North America, thrusting up the Sierra Nevada, the ranges of Nevada and Utah, and then the Rockies as the continent slid westward over the Pacific Plate. Western Colorado bowed and buckled it is true, and was elevated many thousands of feet, but still the sedimentary rock layers remain nearly horizontal.

254

HOGBACK

BUTTE — resistant caprock layer

CUESTA

MESA

This is a colorful land in which many of the sedimentary layers are tinted with shades of salmon, pink, or red, and contrast sharply with the blue of desert sky and the soft greens of pinyon, juniper, and sage. It is a desert land now as it often was in the past, with average rainfall below 10 inches annually. The warm-hued rocks are often bare of soil, and vegetation is usually scanty, making it a geologist's paradise.

All this area is within the drainage basin of the Colorado River, whose tributaries carve its innumerable canyons. Simple folds and faults sometimes control the drainage, blocking established stream routes and forcing rivers to detour or to cut deep into hard, intractable Precambrian rock. These same faults and folds give different elevations to different portions of the plateau country: the Uncompahgre Plateau is 9000 feet above sea level, Mesa Verde is 7000 feet, and the Roan Plateau averages 8000 feet. None have the intense folding and large-displacement faulting that characterize the Rocky Mountain ranges. Three areas now surfaced with volcanic rocks — the White River Plateau, West Elk Mountains, and San Juan Range — are transitional between mountain and plateau, and are treated separately in Chapter V.

The northern part of the plateau area is not as strongly canyoned as the southern, but is warped instead into basins and rolling uplands, sage-covered home of pronghorn antelope. Although surfaced with much younger sedimentary rocks than the rest of the Plateau Province, it has many basic similarities, notably flat-lying rock layers only locally faulted or folded.

Precambrian rocks are exposed in the Plateau Province only in the hearts of the deepest canyons and in a few hills raised by faulting. In southwest Colorado the Precambrian rocks were part of ancient Uncompahgria, an island range lifted high by faulting in Pennsylvanian time as part of the Ancestral Rockies, and then leveled by erosion. Seas that originally surrounded this range retreated westward by the beginning of Triassic time, leaving a wide, flat coastal plain on which rivers deposited red sediments in floodplain and lake and delta. Occasionally dunes of golden sand drifted across the coastal plain, finally to be preserved as thick cross-bedded sandstone layers.

As mountains began to rise in California during Jurassic time, windblown sands became more widespread. Coastal dunes became dunes of a desert interior. Dune sands accumulated over a remarkably wide area stretching from Colorado halfway across Utah to Zion National Park, in a sand sea reminiscent of the sands of Arabia or the Sahara.

Later the vast sandy plains were swept by meandering rivers, and became lush and sometimes even swampy. In the mud and sand deposited at this time, now the Morrison Formation, more than 70 species of dinosaurs found a special kind of immortality — they became fossils.

As the Cretaceous Period began, seas approached again, this time from north and east and south. Soon western North America was experiencing its greatest submergence since before the Ancestral Rocky Mountain uplift. In almost all of Colorado the sandy shore deposits that would become the Dakota Sandstone were blanketed with marine shales that contained the shells and skeletons of innumerable marine animals: coiled ammonites, giant oysters, clams, and swimming reptiles. In the plateau country this is the Mancos Shale, a western Colorado equivalent of the Pierre Shale. Then the sea withdrew and along its margins sandy beaches, muddy swamps, and river deltas left their mark in layers of sand, coal, and mud that form the Mesaverde Group. Here some of the last of the dinosaurs were preserved.

Finally an eastward-moving orogeny, the mountain-building unrest which you remember started in Jurassic time in California, reached into Colorado. The Rockies bowed upward into folded, faulted ranges, their bases close to sea level, their summits two miles high. But the Plateau Province somehow escaped much of the disturbance. Between the Wasatch Range in Utah and the Rockies in Colorado, Mesozoic sediments remained essentially flat-lying. Great rivers fed from

256

The slopes of Mt. Garfield, near Palisade, are sculptured in Mancos Shale. The mesa is capped with Mesaverde Group sandstone. Both formations display the barren cliff-slope-cliff topography characteristic of the western plateau country. JACK RATHBONE PHOTO

Rocky Mountain highlands deposited thick shale layers and cut canyons in the soft new sediments, sometimes through gently warped or faulted structures that were hardly more than ripples on the foreland of the Rockies.

During and since Tertiary time, both deposition and erosion have taken place, with erosion the final sculptor. When the Rockies blocked eastward drainage and the Uinta Mountains blocked northward drainage, an immense lake formed in the area where Colorado, Utah, and Wyoming join, a lake geologists call Lake Uinta. Thick deposits — thousands of feet — of gray silty or sandy mud washed from the rising Rockies down onto vast river floodplains and deltas flanking the lake, to form the Wasatch Formation. In the lake were deposited muds that later became the Green River Shale. Seasonal changes in sediment color and grain size marked the layers of mud deposited in Lake Uinta. Like tree rings the thin annual layers, called **varves**, can be counted, and they show that Lake Uinta lasted 6,500,000 years.

Then for many millenia volcanoes ruled, and lava flows spread from large volcanic centers in Colorado, while drifting

257

ash covered much of the western part of the state. As volcanic activity decreased, uplift began again, with broad regional lifting of Colorado and parts of adjacent states climaxing in Miocene and Pliocene time. Boosted by this renewed uplift, as well as by increased rainfall and snowfall during the Ice Ages, the Colorado River and its tributaries assailed the plateau country, carving intricate canyons, modeling mesas and cuestas and occasional badlands, and blending the scenery of western Colorado with that of adjacent parts of Utah, Arizona, and New Mexico.

The plateau country contains a wealth of important energy resources — oil and gas in Paleozoic limestones, nuclear fuels in Triassic and Jurassic sandstones and shales, coal in Cretaceous strata, and oil shale — the greatest undeveloped fossil fuel source in the world — in the Tertiary Green River Shale. Development of the oil shale rests on difficulties in finding production methods that are at once economic and compatible with environmental concerns. Limitations in the amount of water available in this dry desert land, where river flow is long since spoken for, pose additional problems, for water is needed for processing the oil shale.

interstate 70

rifle — utah line

(88 miles)

West of Rifle, Interstate 70 (U.S. 6) curves southwestward in a big S following the valley of the Colorado River. The river at one time was called the Grand, and from that name we get place names like Grand Valley, Grand Junction, and Grand Mesa.

Roan cliffs north of the highway and the cliffs that edge Battlement Mesa and Grand Mesa to the south expose flat-lying, pinkish gray, limy Tertiary sandstone and shale of the Wasatch and Green River Formations. The Green River Formation, at the top of Roan Cliffs, contains the richest oil-shale beds in the world — more than 1.8 **trillion** barrels of estimated oil reserves, firmly locked into the shale.

I-70
Rifle to Utah line

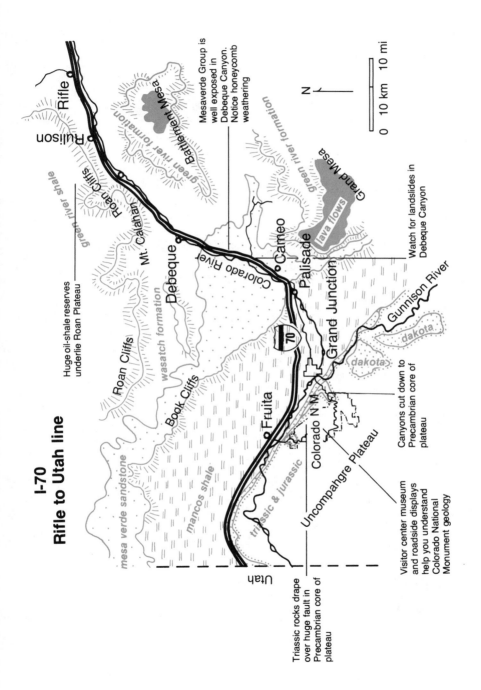

Rifle

Rulison

Rulison

Roan Cliffs

green river shale

Mt. Callahan

green river formation

Battlement Mesa

Mesaverde Group is well exposed in Debeque Canyon. Notice honeycomb weathering

Debeque

wasatch formation

Roan Cliffs

Book Cliffs

Colorado River

Cameo

Palisade

Grand Mesa

lava flows

green river formation

Watch for landslides in Debeque Canyon

Huge oil-shale reserves underlie Roan Plateau

mesa verde sandstone

mancos shale

Fruita

Colorado N M

Grand Junction

dakota

dakota

Gunnison River

triassic & jurassic

Uncompahgre Plateau

Canyons cut down to Precambrian core of plateau

Visitor center museum and roadside displays help you understand Colorado National Monument geology

Utah

Triassic rocks drape over huge fault in Precambrian core of plateau

N

0   10 km   10 mi

259

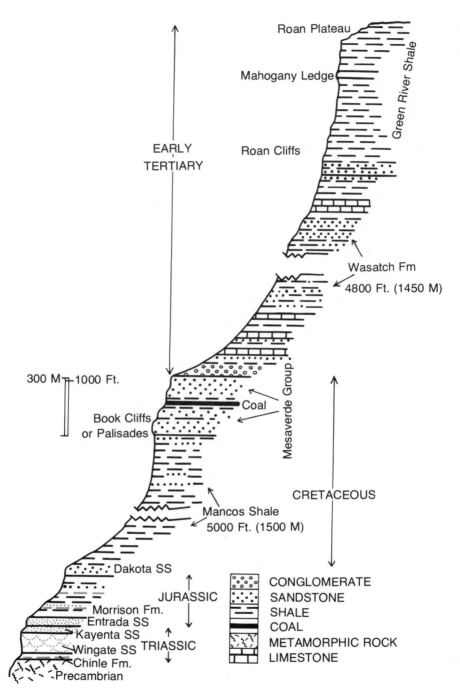

Roan Plateau

Mahogany Ledge

Green River Shale

EARLY
TERTIARY

Roan Cliffs

Wasatch Fm
4800 Ft. (1450 M)

300 M─┬─1000 Ft.

Coal

Mesaverde Group

Book Cliffs
or Palisades

CRETACEOUS

Mancos Shale
5000 Ft. (1500 M)

Dakota SS

JURASSIC

Morrison Fm.

Entrada SS

Kayenta SS

TRIASSIC

Wingate SS

Chinle Fm.

Precambrian

⊙∘∘⊙	CONGLOMERATE
∴∴∴	SANDSTONE
—	SHALE
▬	COAL
〰	METAMORPHIC ROCK
⊞	LIMESTONE

Stratigraphic section between Rifle and Grand Junction

260

Mt. Callahan, rising 3600 feet above the Colorado River near Grand Valley, is capped by basalt flows probably once continuous with those on Battlement and Grand Mesas. These flows help to measure the cutting power of the river, for during the 10 million years since the flows formed, the river has cut 3600 feet downward.

JACK RATHBONE PHOTO

From I-70 you can see the dark brown line called Mahogany Ledge, which averages more than 27 gallons of oil per ton. But it's hard to get out. The search continues for a recovery method that is both economical and environmentally sound. Because oil-bearing parts of the Green River shale thicken and dip northward, most oil shale prospects are north of here under the Roan Plateau.

Most of the downcutting along the Colorado River probably occurred during the canyon-cutting episode that accompanied and followed Miocene-Pliocene regional uplift, when streams were newly steepened and strengthened. But much erosion also must have occurred in Ice-Age time when runoff and stream load were far greater than they are today. Terrace levels in the valley of the Colorado River indicate stages at which the river stabilized for a time and widened its floodplain before beginning another cycle of erosion.

West of the town of Grand Valley the highway descends gradually through the lowest part of the Wasatch Formation. Then in the

261

canyon below Debeque it drops through stairstep cliffs of the Cretaceous Mesaverde Group, shoreline sands deposited on beaches and bars that moved slowly westward and then eastward again as the Cretaceous sea swept over the land and later retreated. Shale and coal layers in the Mesaverde show that muddy or swampy lagoons often festooned the shore; the Mesaverde Group can usually be recognized by their presence. Occasional small springs and seeps are surrounded by bitter white salt deposits brought to the surface by spring waters. Landslides are frequent along the river, and toppled blocks of sandstone form chaotic debris below freshly scarred cliffs, for erosion of soft shale layers in and below the Mesaverde Group undermines the sandstone layers. In flood, the river has in times past been able to remove the landslide material, but now dams hold back flood waters and impair this necessary function.

The Colorado River enters Debeque Canyon through flat-lying sandstone and shale layers of the Mesaverde Group.
JACK RATHBONE PHOTO

Grand Valley Diversion Dam near Mile 50 removes Colorado River water for irrigation. At Mile 44 a large mass of rocky rubble marks a 1950 slide that destroyed a tunnel carrying vital irrigation water from the dam to orchards near Grand Junction. In an all-out effort the tunnel was rebuilt in 19 days, in time to save both crop and fruit trees.

262

Small deep hollows carved by wind, called honeycomb weathering, frequently characterize Mesaverde Group sandstones.

JACK RATHBONE PHOTO

Several Mesaverde coal seams are thick enough to mine. Cameo Mine at Mile 46 produces coal for the power plant across the river. Farther north the coal seams thicken to as much as 50 feet and form one of the largest energy reserves in the country. The coal has a low sulphur content. Never having been squeezed and compressed as were many coal beds in the Appalachians, it is soft or **bituminous** coal.

At the mouth of Debeque Canyon the highway enters the lower part of Grand Valley. Here it is well below the base of the Mesaverde Group, which caps the towering palisades of Mancos Shale called Book Cliffs. This gray shale, yellow where it is leached, contains types of clay that swell when wet and shrink when dry. Such action brings about a loose soil that is so constantly eroding that it won't support much in the way of vegetation. Where it is not protected by the Mesaverde caprock, the Mancos Shale erodes into hump-backed gray and yellow badlands.

Grand Junction lies at the confluence of the Colorado and Gunnison Rivers. The whole of this part of Grand Valley is surfaced with Mancos Shale rendered fertile by irrigation. Because of the shale's fine texture and high clay content the valley has no usable shallow groundwater. Water is obtained from the river or from deep wells drilled down into Jurassic sandstone — the same sandstone that appears on the plateau across the valley. And because the water

263

enters the Jurassic aquifer at a higher point than where the wells are drilled, and is unable to escape through impermeable shale above, it is under hydrostatic pressure and rises in wells without pumping — to a degree. The artesian water supply has been over-developed, and some of the wells that used to flow now have to be pumped.

Across the river to the west rises the Uncompahgre Plateau, a broad, gentle faulted anticline superimposed on ancient Uncompahgria. Deeply dissected pink cliffs of Triassic sandstone directly across from Grand Junction are in Colorado National Monument. Features of the Monument can best be seen by taking the loop road through it, rejoining I-70 at Fruita (see next itinerary). However some of the main features can be seen from I-70.

The present Uncompahgre Plateau probably did not begin to rise until regional uplift in Miocene-Pliocene time. Then it was pushed up more rapidly than surrounding areas, right in the way of the southwest-trending drainage of the Colorado River. The Colorado, augmented by its tributary the Gunnison, now detours around the north end of the uplift. For a time one or both of these rivers probably flowed through Unaweep Canyon, a sizeable gorge that cuts right across the Uncompahgre Plateau (see Co. 141 WHITEWATER to NATURITA).

I-70 detours around the north end of the Uncompahgre Plateau, tracing the boundary between Mancos Shale and underlying Dakota Sandstone. The Colorado River detours too, though it carves a sharp notch into Jurassic and Cretaceous sediments of the plateau at Westwater Canyon near the Utah border. Hogbacks and cuestas of Dakota Sandstone edge the plateau here, with colorful red and green shale and white sandstone of the Morrison Formation often showing in clefts between cuestas, as at Mile 18. Across Grand Valley to the northeast, Book Cliffs curve westward into Utah. Above them, far in the distance, are the gray Tertiary rocks of Roan Cliffs.

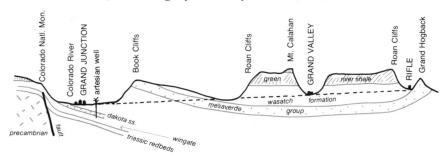

Section parallel to I-70 from Rifle to Colorado National Monument

Massive Wingate Sandstone of Colorado National Monument bends down over the east edge of the Uncompahgre Plateau. Capped by erosion-resistant beds of the Kayenta Formation, these cliffs add color and beauty to the Monument area. Book Cliffs and above them Roan Cliffs rise across the valley of the Colorado River.

colorado national monument

(28-mile loop from grand junction to fruita)

Deep canyons, sheer shadowed cliffs, bright sun, and pale, flesh-colored sandstone streaked with time characterize Colorado National Monument's unusual scenery. The landscape here is Plateau Province landscape, with warm-hued rocks in horizontal layers, with dry stream beds that little reveal the forces of erosion that shaped this land.

Entering the Monument from either end, the road crosses a Dakota Sandstone cuesta and bands of rainbow-hued Morrison shale and red Triassic sandstone and shale. Then abruptly it enters rocks of Precambrian age — dark reddish gray gneiss and schist in canyon bottoms and on scrub-covered slopes below the high pink cliffs. The Precambrian rock is like that in Black Canyon of the Gunnison or in

265

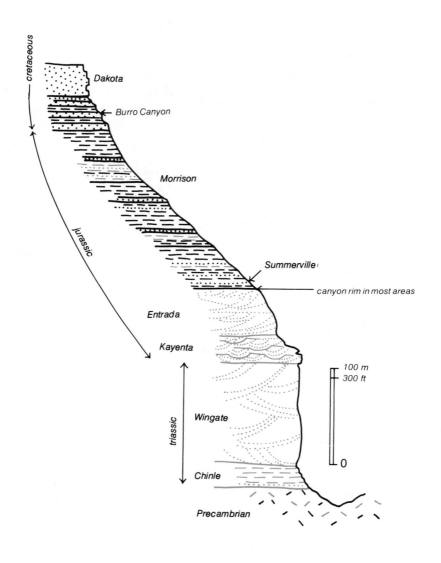

Geologic units in cliffs of Colorado National Monument

the cores of many of Colorado's mountain ranges (and probably like that under most of the continent) — hard, resistant, very old, cut by light veins of granite. Once or many times this rock has formed the roots of ancient mountain systems. It has been broken, folded, partly melted and recrystallized so often that little can be told today about the ancient lands of which it was a part.

The upper surface of the Precambrian rock is an almost perfectly beveled plain. It represents an immensely long period of erosion, a billion years when much of this continent was being trimmed down to sea level.

Sediments of the Paleozoic Era (Age of Fishes) were deposited on this surface, nearly 2000 feet of marine sandstone, shale, and limestone. Then, after the middle of Pennsylvanian time, when a large part of southwestern Colorado became part of the highland geologists call Uncompahgria, they were all washed off again.

On the newly denuded Precambrian surface, which though faulted in places is thought to correspond fairly closely with the ancient peneplain, Mesozoic sediments accumulated. In this region these were mostly continental sediments, for the seas had withdrawn from the land. Red sand and mud were laid down in alternating layers on a broad delta similar to the Nile Delta or a great floodplain like that of the Mississippi. Later, sand was blown in on desert winds — a creeping sea of sand filling the broad valley. Ultimately, as the land sank slowly and evenly, the sea advanced once more, and little by little a beach moved across the desert sands.

The red Triassic sandstone and shale — the Chinle Formation — lies on the denuded Precambrian surface at the base of the immense cliffs. The tall clifs are Triassic, too — the Wingate Sandstone. Notice the cross bedding in this sandstone, long curving lines that indicate, along with other evidence, that it was deposited as dune sand.

Near the top of the great cliffs, ledges back away from the precipices. The lowest of these ledges is the Kayenta Formation. Long after the rock itself was formed, the Kayenta was hardened by silica deposited by groundwater, and so it frequently forms a protective cap on cliffs and monoliths. It is a floodplain deposit, full of scoured out and refilled channels and steep torrent-caused diagonal laminations. The smooth rounded salmon-pink cliff above it is the Entrada Sandstone, a unit that can be recognized all over southwest Colorado and southeast Utah.

The Wingate Sandstone is one of several wind-deposited formations in the plateau area of western Colorado. Broad, swooping across laminations were originally the lee slopes of sand dunes. S.W. LOHMAN PHOTO, COURTESY OF USGS

Back from the rim in the south part of the Monument, colorful shales of the Jurassic Morrison Formation lie above the Entrada. Recently the world's largest dinosaur skeleton, remains of a monster as tall as a six-story building, were found in these rocks outside the Monument boundary. Above the Morrison Formation, Cretaceous Dakota Sandstone forms the plateau surface. Near the north end of the Monument, both the Dakota and Morrison have been washed away, and the upper surface of the plateau is on the Entrada Sandstone.

Look eastward to Book Cliffs and notice that sandstones of the Mesaverde Group (also Cretaceous) which cap them are at approximately the same level as the surface of the Entrada Sandstone here. This will give you a measure of the uplift of Colorado National Monument and the Uncompahgre Plateau, for there are 600 feet of Morrison Shale, 130 of Dakota Sandstone, 5000 feet of Mancos Shale, and about 1000 feet of Mesaverde Sandstone above the Entrada. Added together these figures show that total uplift of the Uncompahgre Plateau relative to Book Cliffs is in the neighborhood of 6700 feet.

That the Uncompahgre Plateau has been lifted relative to its surroundings is undeniable, but **when** it was lifted is hard to say. Some geologists believe that it rose in very early Tertiary time, along

268

with the Rockies to the east, and then again in Miocene-Pliocene time when regional uplift took place. Others feel that the plateau was not elevated in early Tertiary time at all, and they cite as evidence a canyon that cuts deeply through the present Uncompahgre Plateau. A big river like the Colorado or the Gunnison does seem to have flowed southward directly across the area, and presumably as the plateau rose in Miocene-Pliocene time the river first carved and then abandoned this gorge (see CO. 145 WHITEWATER to NATURITA), leaving behind intriguing but incomplete clues to the history of the area.

Erosion has doubtless removed thousands of feet of poorly consolidated sedimentary rocks from the plateau surface — Cretaceous and Tertiary formations to correspond with those visible in Book Cliffs and Roan Cliffs. As the Colorado, the Gunnison, and the Uncompahgre Rivers deepened Grand Valley, tributaries cut back into soft Mesozoic sediments of the Uncompahgre Plateau and undercut the massive Wingate Sandstone to form the spectacular cliffs and precipices of Colorado National Monument.

The Uncompahgre Plateau is a nearly flat-topped highland about 25 miles wide and 100 miles long. The Precambrian rocks which form the plateau core are faulted at both its edges. In many places the sedimentary rocks above them drape over the faults, bending rather than breaking. Such draped sediments are exposed where the road begins its descent back into Grand Valley at either the north or south end of the National Monument.

Jurassic and Triassic formations bend in a monocline across the great fault that edges the Uncompahgre Plateau.

S.W. LOHMAN PHOTO, COURTESY OF USGS

269

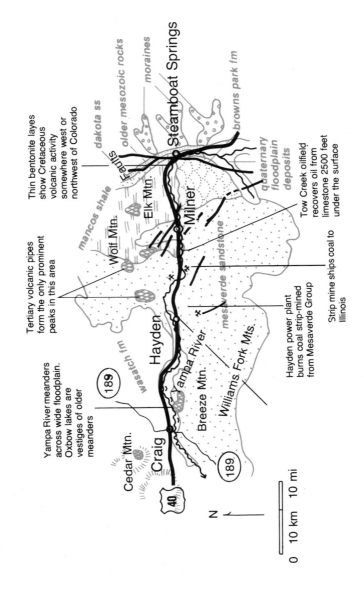

Thin bentonite layes show Cretaceous volcanic activity somewhere west or northwest of Colorado

Tertiary volcanic pipes form the only prominent peaks in this area

Yampa River meanders across wide floodplain. Oxbow lakes are vestiges of older meanders

Tow Creek oilfield recovers oil from limestone 2500 feet under the surface

Strip mine ships coal to Illinois

Hayden power plant burns coal strip-mined from Mesaverde Group

dakota ss

older mesozoic rocks

moraines

browns park fm

mancos shale

Wolf Mtn.

Elk Mtn.

Steamboat Springs

Milner

quaternary floodplain deposits

mesaverde sandstone

faults

wasatch fm

Hayden

Yampa River

Breeze Mtn.

Williams Fork Mts.

189

Cedar Mtn.

Craig

189

40

N

0   10 km   10 mi

u.s. 40
steamboat springs — craig
(43 miles)

Several of the hot springs at Steamboat Springs bubble up in the city park west of the main part of town; the swimming pool at the east end of town is supplied by others. Water in these springs, heated deep below the surface, comes up rapidly along a fault zone into the Dakota Sandstone, from which it issues. The springs have left thick gray travertine deposits all along the base of the high hill south of town, a hill formed of Tertiary deposits of dune sand and stream gravel.

The valley west of Steamboat is underlain by Mancos Shale dipping gently westward off the west flank of the Park Range. This fine gray Cretaceous marine shale contains occasional fossils, and in roadcuts and along the river banks you can see that it also has thin orange seams, bentonite layers formed by gradual chemical change in layers of volcanic ash. The Mancos Shale is equivalent in position and roughly equivalent in age to the Pierre Shale of eastern Colorado.

A dark peak north of the highway is Elk Mountain, a vertical pipe of igneous rock that may never have reached the surface. A number of similar intrusions occur west of it, but can't be distinguished from the road.

Just east of Milner the lower part of the Mesaverde Group (Cretaceous) forms low ridges north and south of the valley. The Mesaverde Group, once called the Mesaverde Formation, is now divided into three formations, with the old name still retained for the group of three. Formed on Cretaceous shores, this sandstone-shale complex is widespread in western Colorado and adjacent states because Cretaceous seas, which once covered the whole state, retreated eastward slowly forming these shore deposits as it went. The top of the Mancos Shale here is older than the top of the Pierre Shale east of the mountains, and the Mesaverde is correspondingly slightly older than the Upper Cretaceous sandstone in the eastern part of the state. The age differences show that the retreat of the sea was gradual and fluctuating, and that millions of years passed before it had receded from the entire state.

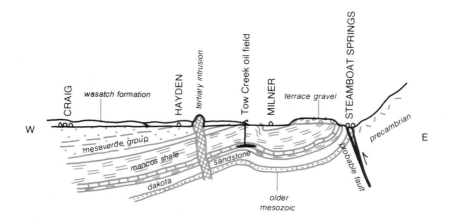

Section along U.S. 40 between Steamboat Springs and Craig.

Mesaverde sandstones are fairly easy to recognize. Notice their tan color (though often they are covered with black lichens) and their tendency to develop deep fist-sized cavities as they weather, a feature called **honeycomb weathering**. They also tend to form large caves like the famous ones sheltering cliff dwellings in Mesa Verde National Park. Like Cretaceous sandstone east of the mountains the Mesaverde contains numerous coal layers, some of them quite thick, formed in swamps and lagoons similar to the swamps and lagoons that border Atlantic and Gulf Coast states now. Quite a few coal mines are now operating in western Colorado, most of them open-pit mines recovering coal from seams three to ten feet thick. The coal is used locally to generate electricity, or shipped to midwestern markets.

A little west of Milner the highway crosses a prominent anticline. Watch the directions of dip in highway cuts to spot the crest of the anticline, where rocks that are dipping east level off and then begin to dip west. Coal is mined along this anticline, and oil wells at its crest (difficult to see, but you can spot small storage tanks) take advantage of oil's tendency to rise through porous rock. Here, the permeable rock is fractured Cretaceous limestone about 2500 feet below the surface, underneath the Mancos Shale. As the limestone arches across the anticline, it forms a natural trap for oil, which can't escape through overlying impermeable shale. Some deeper wells in this field reached the Dakota Sandstone, but it yielded hot water instead of oil! As at Steamboat Springs, temperatures in this structure must increase unusually rapidly with depth.

On the west side of the anticline the highway crosses younger and younger (higher and higher) parts of the Mesaverde Group. Slope-

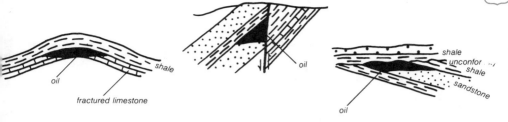

Unable to escape upward through impermeable shale, oil and gas collect in upside-down pools in a variety of geologic "traps."

forming shales alternate with the more typical cliff-forming sandstone and coal. More coal mines appear, many of them on sites of long-abandoned underground mines. All of the present open-pit mines are required by law to backfill and revegetate spent open-pit areas.

The double peak on the skyline far to the north, Bears Ears, is a landmark in the volcanic Elk Head Mountains. To the south, in the Flattops area, horizontal lava flows top the White River Uplift, a dome-shaped anticline that is a westward extension of the Rocky Mountains, a sort of transition between eastern faulted anticline ranges and flat-topped western plateaus.

A big power plant near Hayden uses local coal from a mine a few miles south. This town was named for Ferdinand V. Hayden, a pioneer geologist who explored and mapped western Colorado in 1869 and 1870. His *Geological and Geographical Atlas of Colorado* was published in 1877, and still makes fascinating reading.

West of Hayden, flat-lying light brown shale and sandstone begin to appear above the Mesaverde Group. The lower 250 feet or so of these rocks are marine Upper Cretaceous strata, but upper parts contain Tertiary plant fossils and are part of the non-marine Wasatch Formation. At Craig, they are more than a thousand feet thick. Rounded, knobby lumps in these sandstones formed **after** the rock was deposited, when groundwater rich in minerals flowed through the sediments and deposited some of its minerals around small nuclei such as little bits of wood, decaying leaves, or fossil shells. Once the **concretions**, as they are called, started to form, more minerals adhered to them, and they grew layer by layer to the size you see here.

Breeze Mountain southeast of Craig is another Tertiary intrusion. This one has lifted surrounding sedimentary rocks, including rocks of the Mesaverde Group, into a dome known by geologists as Breeze Anticline or Craig Dome. Farther southwest are the Williams Fork Mountains.

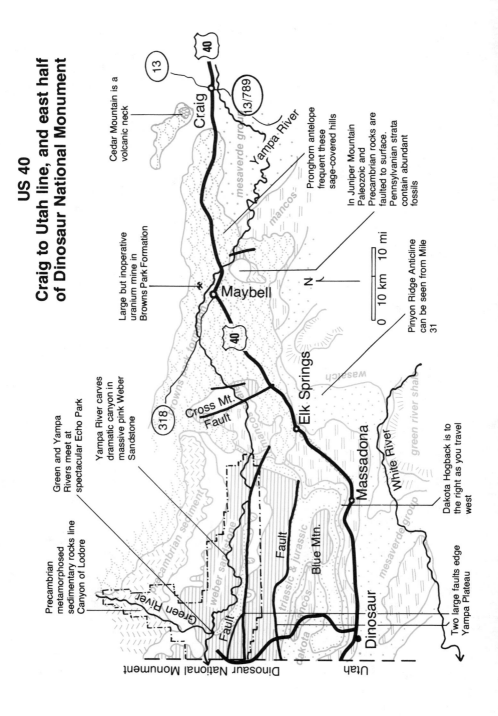

US 40
Craig to Utah line, and east half
of Dinosaur National Monument

40

40

13

13/789

Craig

Cedar Mountain is a volcanic neck

mesaverde group

Yampa River

mancos

Pronghorn antelope frequent these sage-covered hills

In Juniper Mountain Paleozoic and Precambrian rocks are faulted to surface. Pennsylvanian strata contain abundant fossils

Large but inoperative uranium mine in Browns Park Formation

Maybell

N

0   10 km   10 mi

Pinyon Ridge Anticline can be seen from Mile 31

40

Elk Springs

wasatch

318

Cross Mt. Fault

Massadona

White River

green river shale

mesaverde group

Dakota Hogback is to the right as you travel west

Green and Yampa Rivers meet at spectacular Echo Park

Yampa River carves dramatic canyon in massive pink Weber Sandstone

Fault

Blue Mtn.

triassic & jurassic

mancos

dakota

Dinosaur

Precambrian metamorphosed sedimentary rocks line Canyon of Lodore

Green River

precambrian sediment

weber sandstone

Fault

Two large faults edge Yampa Plateau

Dinosaur National Monument

Utah

u.s. 40
craig — utah line

(90 miles)

Craig is close to the Yampa River, surrounded by Cretaceous sedimentary rocks: Mancos Shale flooring the river valley and Mesaverde Group sandstone and shale in the hills to north and south. A light-colored Tertiary volcanic plug forms Cedar Mountain northwest of town.

Looking back at the Park Range you can see the horizontal surface of the Tertiary pediment quite well. North and Middle Parks, as well as all this area west of the Park Range, are thought to have been filled nearly to that level with Tertiary deposits and volcanic debris. Following Miocene-Pliocene uplift less than 28 million years ago, revitalized streams scoured away vast quantities of these poorly consolidated deposits and shaped today's topography.

An open pit mine visible southwest of Craig obtains coal from Mesaverde Group coal seams. The nearby power plant was to produce electricity for towns to the south, but conflicting water rights have at the time this is written delayed its operation.

About nine miles west of Craig, U.S. 50 climbs onto Tertiary sandstone — the Browns Park Formation, white cross-bedded sandstone and yellowish siltstone. These rocks contain a high proportion of volcanic ash, suggesting that at least some of the volcanic plugs in this region may have reached the surface. They also contain uranium. There are large volcanic fields south of here in the Flattops area and still farther south in the West Elk Mountains; some of the ash may have come from there.

The Browns Park Formation is often capped with a **desert pavement** of rounded pebbles and cobbles which have been "let down" from higher parts of the formation or from Quaternary stream deposits. Finer parts of the formation, not as well protected by vegetation as they might be in a moister climate, have blown away. The cross bedding in the Browns Park Formation indicates that parts of this rock unit were laid down as dunes; bands of pebbles and clay suggest that other parts were stream deposits.

The low hilly skyline south of the highway is punctuated about 25 miles west of Craig by Juniper Mountain, a little cluster of higher hills made of Paleozoic and Precambrian rocks jutting up through the sea of the Browns Park Formation. They seem to be a small fault block but details of the faulting are covered by surrounding Tertiary sediments. The Yampa River flows right through the little range. Upstream from these mountains the river follows a meandering course across a broad floodplain that may have formed because the hard rocks of Juniper Mountain acted as a partial dam.

Notice the prominent cross bedding in the Browns Park Formation west of Maybell. Some of the curving laminae show that this part of the Browns Park Formation was once a sand dune area. Being poorly consolidated (as are most Tertiary formations, for it takes time to harden sedimentary rock), some of the sandstone now is turning back into dunes.

At Mile 62, Cross Mountain shows to the northwest. This small range, a north-south faulted anticline, is quite similar in structure though not in size to the larger faulted anticlines that make up the main ranges of the Colorado Rockies. Paleozoic rocks still arch across most of Cross Mountain. The Yampa River does some remarkable things here, boldly cutting right through the Paleozoic sediments and into Precambrian granite of Cross Mountain, and then continuing westward to tackle the Yampa Plateau at the eastern end of the Uinta Uplift. Geologists believe that the course of the Yampa was established while the whole region was submerged in Tertiary sediments, and that as the land rose during Miocene-Pliocene uplift the river cut down through these deposits into older rocks so gradually that it generally adhered to its original course. It is probably pure coincidence that it tackles Juniper Mountain, Cross Mountain, and the Uinta Uplift instead of taking the easier course around them.

As you approach Cross Mountain notice the gray cliff of Mississippian Leadville Limestone arching across the fault block of Precambrian granite that is the core of the mountain. At the fault on the west side, it butts right into Cretaceous rocks.

Lower Tertiary sediments form hills to the south. These gray shales and sandstones, the Wasatch and Green River Formations, are older than the Browns Park Formation. The Wasatch sandstones and shales, 2500 feet of them, are sediments washed off the rising Rockies in Paleocene time. The Green River Shale accumulated in ancient Lake Uintah, a lake that flooded much of western Colorado, eastern

276

Utah, and southern Wyoming in Eocene time. Fossil fish, crocodiles, insects, and plants occur in the Green River Shale in Colorado. Equivalent strata north of here near Fossil, Wyoming, now in Fossil Butte National Monument, are famous for their abundant and marvelously preserved fossil fish. Upper parts of these gray shales contain an oil-like substance called kerogen, and form the Green River Oil Shale, one of the greatest energy reserves in the entire world.

The jumbled country of the eastern part of Dinosaur National Monument now appears to the west, with the V-shaped notch of the Yampa Canyon visible at its heart. This canyon is cut in thick, massive, salmon-colored Weber Sandstone which, though it contains no fossils, is clearly Pennsylvanian and Permian in age because it lies above known Pennsylvanian marine shales and below known Permian and Triassic rocks. Its broad cross bedding shows that parts of it at least are dune deposits; red shales between the dune-like layers suggest pond or interdune or lake deposits or possibly deposits on an ancient desert delta.

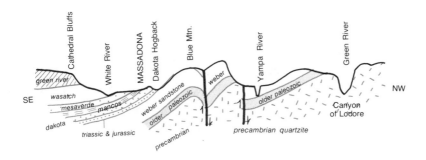

Section across U.S. 40 south of and into Dinosaur National Monument

At Elk Springs the highway leaves the Browns Park Formation and drops into Cretaceous Mancos Shale again. The geology is much simpler than it looks: Blue Mountain between the highway and the higher Yampa Plateau is an anticline edged by upturned Mesozoic rocks. At Mile 27 the highway parallels the curving edge of the Dakota Hogback, which encircles the east end of the Blue Mountain anticline.

Keep in mind that from here all the way to the Utah line the highway will be on Mancos Shale. The Dakota Hogback will form the ridge just north of the highway, and older rocks will lie to the north, younger ones to the south. Look up the little stream valleys that cut

277

On the south flank of Blue Mountain, upturned Mesozoic sedimentary rocks form rows of angular flatirons. The same rocks form the horizontal bands on top of the mountain.

W.R. HANSEN PHOTO, COURTESY OF USGS

through the Dakota Hogback to see the colorful green and purple shales of the Morrison Formation, and beyond them deep vermilion Triassic shale and sandstone. Pale pink mountain flanks are Weber Sandstone.

West of Massadona, hogbacks in the Mesaverde Group (to the south) can be seen extending west toward Utah. Notice the way geology controls topography, and even vegetation, here. Hogbacks result from differential erosion, with soft shaly and coaly layers eroding away more easily than hard sandstone layers. Because of the lack of rainfall in this arid western country, a true desert deprived of Pacific moisture by mountains to the west, erosional processes are governed more by wind than by water. You have already seen the desert pavement of stones concentrated on top of the Browns Park Formation, and the dunes forming from sand from that formation. Wind attacking soft and hard rock layers writes geology in bold type across the land. It is helped of course by occasional rain and flood, but many of the mini-valleys between these rows of hogbacks show no sign of stream channels. Assisted on steep slopes by gravity, wind has cleaned off the Weber Sandstone surfaces of Blue Mountain and the Yampa Canyon country, as you will see if you take a trip into eastern Dinosaur National Monument (itinerary follows).

278

From the town of Dinosaur, where streets are named for ancient reptiles, look north to the dramatic face of Roundtop Mountain. Dakota Sandstone flexes sharply up its flank and across its summit. Right at the Utah line you can see the rainbow hues of Jurassic and Triassic rocks between sculptured slopes of the Dakota Sandstone. Farther west, pink sandstone of the Weber Formation arches up onto Split Mountain in the Utah part of Dinosaur National Monument.

dinosaur national monument,
east half

(46 -mile round trip from u.s. 40)

The road into Dinosaur National Monument, branching from U.S. 40 two miles east of the town of Dinosaur, rises onto the Yampa Plateau and reaches the brink of the deep canyon cut by the Yampa

The canyon of the Yampa River gashes the surface of the Weber Sandstone. Its winding course is thought to date back to a time when the river flowed across a vast plain of Tertiary sediments high above its present course. Steamboat Rock, where the Yampa joins the Green River, is at the bottom of the photo.

W.B. CASHION PHOTO, COURTESY OF USGS

279

Precambrian rocks of the Canyon of Lodore are not the granite, gneiss, and schist of the cores of the Rocky Mountain ranges, but sedimentary rock which is scarcely metamorphosed at all. Sandstone, siltstone, and conglomerate are all recognizable, though all are hardened and firmly cemented by time. These old sediments extend north through Wyoming and Montana and into the Canadian Rockies.

W.R. HANSEN PHOTO, COURTESY OF UGS

River. The road first passes through an upturned hogback of Dakota Sandstone and crosses a narrow valley of Jurassic and Triassic sedimentary rocks, vividly colored in shades of green, purple, and red. The fossil quarry for which Dinosaur National Monument is famous is in corresponding Jurassic rocks — the Morrison Formation — about 25 miles west in the Utah section of the park.

The road climbs the plateau on the Dakota Sandstone, but comes onto salmon-colored Weber Sandstone abruptly as it crosses one of the two faults that form the plateau. This is the same Pennsylvanian-Permian rock visible along Blue Mountain. Here you can look more closely at the thick sandstone layers, with their broad, sweeping crossbedding, and at the red shale and siltstone beds that lie between them. These rocks are thought to have been deposited as dunes of an ancient desert, with occasional mudflats wet by sporadic rains.

At the rim of the Yampa Canyon, at the northernmost of the two large faults (at either Echo Park or Iron Springs Bench overlook), look down nearly 3000 feet to the meeting of the Green and Yampa Rivers at Echo Park. (The dirt road to Echo Park is passable by car in dry weather, but quickly becomes impassable if it rains.) This fault adds a thousand feet to the depth of the canyon. Along its cliff-like scarp, side canyons cut into older Paleozoic and then Precambrian rocks. But the cliff face is mostly Weber and the bends in this massive rock, which is usually at least 1000 feet thick, stand out dramatically as evidence of slow but powerful stresses within the earth's crust and the way they can and do bend solid rock.

Down in the canyon north of the fault the Weber Sandstone, often devoid of vegetation or soil, dips gently southward. A small fault separates Iron Springs Bench from the lower cliffs.

The canyon of the Green River also can be seen from these overlooks. This is the Canyon of Lodore, named in 1869 by John Wesley Powell, a Civil War army major who survived the loss of an arm to drift down the Green River with nine companions on a daring boat trip that took them through the wild waters of the Green and Colorado Rivers and through Grand Canyon. Powell later was instrumental in founding the United States Geological Survey, of which he was the second director.

To the west rise the high summits of the Uinta Mountains, a faulted anticline, the only mountain range in Colorado that has an east-west trend. Most of the range is in Utah, though the Yampa Plateau, where you are now, is its eastern end. Colorado's oldest rocks, the Red Creek Quartzite, 2.3 billion years old, are in these mountains.

black canyon of the
gunnison national monument
(56-mile round trip from montrose)

Within the 12-mile stretch set aside as a National Monument, the Black Canyon is narrower and deeper than any other canyon in the country: 1730 to 2425 feet deep and, at the Narrows, 1300 feet from rim to rim. Eroded by the Gunnison River, it is walled with Precambrian gneiss, very hard, very strong rock capable of standing in nearly vertical canyon ramparts. The canyon was carved sometime after the volcanoes in the West Elk Mountains to the north and the

San Juans to the south had ended their long period of eruption. Regional uplift 28 to 10 million years ago had given the river a steep gradient, and during the Ice Age — just in the last two million years — the river was swollen and strengthened with runoff from highland glaciers and loaded with sand and boulders necessary as cutting tools.

Once, the river flowed many miles north of its present course, across land that now underlies the West Elk Mountains. Each time erupting West Elk volcanoes spewed out lavas and expanded the volcanic field, its course swung south, and each time San Juan volcanoes farther south erupted, it was pushed north again. The river must always have sought the lowest route between the two volcanic regions. Finally, as volcanic activity lessened and died, the river cut through the layers of lava and ash, through underlying sedimentary rocks, and into a ridge of tough Precambrian rocks that lay across its course. Here, the river had to work with harder rock, but it slowly scoured a channel into the Precambrian gneiss, pounding and scratching a deeper and deeper and still deeper canyon.

Meanwhile, tributaries stripped the volcanic and sedimentary rocks back from the rim. Draining lower, unglaciated country rather than the high ranges to the east, they were no match for the Gunnison, and once on the hard Precambrian rock they could not cut down as fast. Now they enter the Black Canyon in unusual hanging canyons. Islands and pinnacles within the main canyon were created by weathering and erosion along vertical joints and fissures, with dark shadowy clefts etched into some of the largest joints.

In the shadowy Narrows of the Black Canyon, the Gunnison River, lined with stream-polished boulders, is only 40 feet wide (note figure). Here the canyon is about 1750 feet deep and about 1300 feet from rim to rim.

W.R. HANSEN PHOTO, COURTESY OF USGS

Black Canyon gneiss shows intricate contortions caused by flowage under great heat and pressure in the depths of the earth's crust. Light bands are mostly feldspar and quartz; dark bands are quartz and biotite (black mica). W.R. HANSEN PHOTO, COURTESY OF USGS

Precambrian rocks in the Black Canyon are remarkably similar to some of those exposed in the cores of the Rocky Mountain ranges farther east, as well as to those in the Inner Gorge of Grand Canyon. Indeed they are representative of the rocks that underlie most of the continent. Gneiss, highly distorted and probably compressed and recrystallized deep within the earth's crust nearly two billion years ago (and perhaps before that, too) is interlaced with pink granite and pegmatite dikes squirted into cracks and fissures under tremendous pressure about 1350 million years ago. Look closely at this rock at the many viewpoints, noticing the unusually large crystals in the pegmatite dikes, formed from the watery fluid left over as the granite crystallized. Joint systems determine the course of tributary streams.

No other canyon combines such depth and narrowness, such dark walls and deep shadows, as the Black Canyon. The 2300-foot Painted Wall is the highest cliff in Colorado — nearly twice as high as the Empire State Building. As you can see, little sunlight reaches the canyon depths.

A short distance back from the rim, above the ancient Black Canyon Gneiss, Mesozoic sandstone and shale layers lie on the beveled surface of the Precambrian, above an unconformity that spans more than a billion years. Paleozoic sediments once lay on this surface, but washed away as the Ancestral Rockies were formed and this area rose

283

above the sea as part of the ancient highland of Uncompahgria.

The road from the National Monument descends through Mesozoic sandstone and shale — colorful red and purple and light green rocks. It offers a spectacular view of the jagged north edge of the San Juan Mountains, a volcanic range 40 miles south. Below the San Juans spreads the Uncompahgre Valley, outlined with humpy gray and yellow badlands of Cretaceous Mancos Shale. Westward, resistant sandstone of the Dakota Formation, which crossed the valley underneath the Mancos Shale, rises onto the Uncompahgre Plateau, the forested ridge that fills all the western skyline.

From this point west the terrain is characteristic of the Colorado Plateau Province: flat-lying (or nearly so) rocks mostly of Mesozoic age, eroded by the Colorado River and its tributaries into brightly colored plateaus, mesas, and buttes.

u.s. 50

montrose — grand junction

(61 miles)

Montrose lies on a flat-topped terrace of gravel, sand, and boulders excavated from the San Juan Mountains by glaciers and carried here by the Uncompahgre River when it was swollen with glacial meltwater.

On the western skyline the Uncompahgre Plateau stretches into blueness northward. This forested highland is underlain by a large block of Precambrian rock faulted along both sides and pushed up nearly 7000 feet. Overlying Mesozoic sedimentary rocks are sometimes broken along the same faults, but more commonly, on this side in any case, they stretch and drape over the fault into a monocline, the simplest of folds. Fruitland Mesa and the Black Canyon of the Gunnison, on another upfaulted block, lie to the northeast.

In the distance straight north is Grand Mesa, a large flat-topped plateau between the Colorado and Gunnison Rivers, capped by basalt flows that rest on a thick sequence of Tertiary shale and sandstone — the Green River and Wasatch Formations. Cretaceous rock — the Mesaverde Group — underlies them and forms a small cliff about halfway up the slope. Yellow and gray Mancos Shale, also Cretaceous in age, makes the eroded, barren lowest slopes and surrounding badlands. The Mancos Shale continues across the fertile valley of the

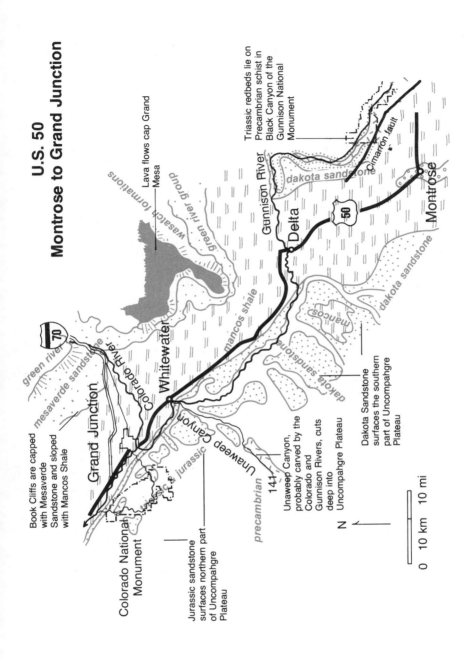

U.S. 50
Montrose to Grand Junction

Book Cliffs are capped with Mesaverde Sandstone and sloped with Mancos Shale

Lava flows cap Grand Mesa

Triassic redbeds lie on Precambrian schist in Black Canyon of the Gunnison National Monument

wasatch formations

green river group

green river

mesaverde sandstone

mancos shale

Colorado River

Gunnison River

Grand Junction

Whitewater

Delta

Montrose

dakota sandstone

Cimarron fault

mancos

dakota sandstone

dakota sandstone

Colorado National Monument

Unaweep Canyon

jurassic

precambrian

Jurassic sandstone surfaces northern part of Uncompahgre Plateau

Unaweep Canyon, probably carved by the Colorado and Gunnison Rivers, cuts deep into Uncompahgre Plateau

Dakota Sandstone surfaces the southern part of Uncompahgre Plateau

N

0   10 km   10 mi

285

Uncompahgre and Gunnison Rivers, and is well exposed close to the highway a few miles north of Delta.

The gently sloping surface between Grand Mesa and the highway is a well developed pediment, an erosion surface that planes across the bedding of the sedimentary rocks. Rapid downcutting by two powerful rivers, the Colorado and the Gunnison, as they churn through soft shale of the valley floor, have made erosion the ruling force in shaping landforms here. The mountains are not surrounded by broad alluvial fans as are those bordering the San Luis or Arkansas Valleys, where deposition and valley-filling are the rule.

In Delta the Uncompahgre and Gunnison rivers come together. The Gunnison has the larger flow, so its name is retained north to Grand Junction, where it joins the Colorado. The Colorado River was called the Grand (hence Grand Junction, Grand Mesa, and Grand Valley) until the State of Colorado persuaded Congress to change the name to Colorado River — the name used in Utah and Arizona — for the great river that has its source in the high ranges of the state.

North of Delta, boulders of basalt are strewn over hills of Mancos Shale, as at Mile 64. These either tumbled from Grand Mesa when it was larger, settled as soft shales were eroded out from under them, or were brought here by mudflows as violent storms swept the area. The basalt is dark reddish or almost black in color, full of little round **vesicles** indicating an abundance of gas bubbles in the molten lava. Small vesicles are filled with white crystals of zeolite. Being a particularly fluid type of volcanic rock, especially when it contains lots of gas, basalt often flows rapidly from cracks and fissures to spread into featureless nearly horizontal landscapes. Many successive thin flows of basalt cap Grand Mesa.

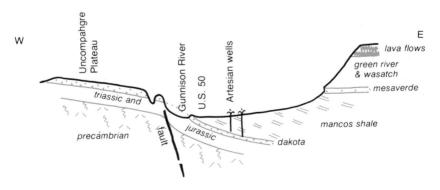

Section across U.S. 50 between Montrose and Grand Junction.

286

The highway gradually approaches the foot of the Uncompahgre Plateau, and north of Mile 54 is close to the Dakota Sandstone cuesta that outlines it. The Dakota rises gently southwestward and then flattens out suddenly to form the plateau surface. You can see its edge where streams have cut into the plateau. Between the canyon rims and the nearby cuesta it has eroded away, revealing pink Jurassic sandstone and shale below it. Sheer cliffs in some of the deeper canyons are in Wingate Sandstone, a Triassic formation made of windblown sands that once were tall desert dunes. Near Grand Junction the sculptured Wingate Formation makes the high cliffs of Colorado National Monument.

Close at hand, limonite geodes form a line of brown pimples in the Mancos Shale east of the road at Miles 51-47 and 45-42. At the heart of each geode is a hollow containing a cluster of calcite crystals, some of them two inches long. The geodes are thought to have formed on the floor of the Cretaceous sea soon after deposition of the mud that is now the Mancos Shale. Temporary heavy iron and calcium carbonate concentration in the seawater must have helped the geodes develop at one level in the formation.

From here you can also see a number of landslides along the edge of Grand Mesa, with soft Tertiary shale contributing both material and slippery-when-wet skid surfaces.

As you near Whitewater, a prominent canyon shows up to the west, carved deeply into Mesozoic rocks of the Uncompahgre Plateau. This is Unaweep Canyon, occupied today by two minuscule streams, East Creek and West Creek, flowing in opposite directions from a nearly indiscernible divide on the floor of the canyon. The canyon slices through 1500 feet of sedimentary rocks, and, in the heart of the plateau, through 1000 feet of Precambrian core, forming a canyon impressive even in this land of canyons (see CO 141 WHITEWATER to NATURITA). It could not possibly have been carved by little East and West Creeks. Geologists agree that a much larger river must once have flowed through it, but they disagree as to whether that river was the Gunnison, the Colorado, or both. Sometime late in Tertiary time uplift of the plateau exceeded the river's ability to cut down through the hard Precambrian rock, and a "pirate" stream flowing northwest through softer sedimentary rocks "captured" the flow of the larger river.

North of Whitewater two sets of cliffs come into view ahead: the lower Book Cliffs of Mancos Shale capped by Mesaverde Group, and the upper Roan Cliffs (edging the Roan Plateau), formed of light gray and grayish pink Tertiary shale and sandstone of the Wasatch and Green River Formations.

Near Grand Junction, in Grand Valley, many farms are irrigated with artesian wells in which water rises without pumping. They tap Jurassic sandstone layers that bend down over the edge of the Uncompahgre Plateau and flatten out under Grand Valley (see I-70 RIFLE to UTAH LINE). Confined beneath impermeable shale, the water flows along the dipping sandstone beds, developing hydrostatic pressure as it drops below the level at which it entered the rock. Overuse of artesian wells, however, has lowered the pressure and many wells that once flowed now have to be pumped. Some irrigation water is brought from the Colorado River for farms and orchards of Grand Valley.

Grand Junction is a major uranium center with U.S. Department of Energy offices and headquarters of many uranium companies. The ore comes from Permian, Triassic, and Jurassic strata that border the plateaus.

u.s. 160

pagosa springs — durango

(55 miles)

Pagosa's hot springs — a large morning-glory pool and some marshy hot seeps — are just south of town across the San Juan River bridge. You can see droopy gray spring deposits on the river bank as you cross the bridge. One of the largest hot springs in America, the big pool discharges 700gallons a minute, at a temperature of 153°F. Hot waters nearly always carry minerals dissolved from rock through which they pass; in most springs the predominant mineral is calcium carbonate derived from limestone. But there is little or no limestone underlying the Pagosa region, so the hot water is charged instead with silica. When the water cools, the silica precipitates as siliceous sinter, another form of quartz. These springs are unusually heavily mineralized, and the sinter deposits around them are massive enough to deflect the course of the San Juan River so that it arcs westward around the springs area. Much of the water is now piped to swimming pools, fountains, and heating systems, and the spring no longer overflows to deposit sinter. The rotten egg odor comes from hydrogen sulfide released from the water as gas.

West of Pagosa Springs, U.S. 160 rises onto a high anticline of Dakota Sandstone, complicated by faults that can be identified by repetition of gray Mancos Shale that normally lies above the Dakota.

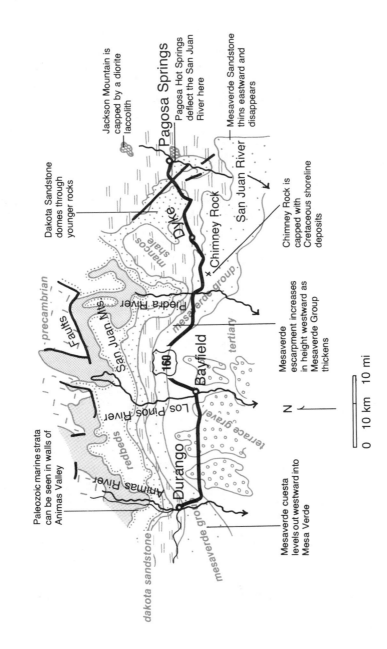

US 160
Pagosa Springs to Durango

Jackson Mountain is capped by a diorite laccolith

Pagosa Springs

Pagosa Hot Springs deflect the San Juan River here

Mesaverde Sandstone thins eastward and disappears

Dakota Sandstone domes through younger rocks

Chimney Rock

San Juan River

Chimney Rock is capped with Cretaceous shoreline deposits

precambrian

Faults

San Juan Mts.

Piedra River

mancos shale

mesaverde group

Bayfield

160

tertiary

Los Pinos River

Mesaverde escarpment increases in height westward as Mesaverde Group thickens

N

terrace gravel

redbeds

Paleozoic marine strata can be seen in walls of Animas Valley

Animas River

Durango

dakota sandstone

mesaverde group

Mesaverde cuesta levels out westward into Mesa Verde

0   10 km   10 mi

289

From the crest of the anticline, near Mile 141, the San Juan Range can be seen to the north. Although this range is mostly volcanic, along its western, northern, and southern edges Paleozoic and Mesozoic sedimentary rocks are tilted up toward a faulted core of Precambrian rock.

Prominent cuestas have now begun to appear to the south. Many of them are capped with sandstone of the Mesaverde Group, hard Cretaceous near-shore sandstone with some shale and coal layers. Sloping sides of these mesas are soft Mancos Shale, no doubt deposited in the same widespread Cretaceous sea as the Pierre Formation east of the Rockies, a sea that stretched from the Arctic to the Gulf of Mexico across North America. These two rock units are important landscape-makers across southwestern Colorado and adjoining parts of New Mexico, Arizona, and Utah.

West of the Dakota-surfaced anticline, the highway gradually moves through the Mancos Shale and the overlying Mesaverde Group sandstones. The top of the Mesaverde Group is near Mile 125, at the base of Chimney Rock, a prominent landmark that rises more than a thousand feet above the surrounding area. The hard cap of Chimney Rock looks like Mesaverde Sandstone, but it isn't. It is Pictured Cliffs Sandstone, above dark gray Lewis Shale. These two formations are easily confused with the Mesaverde-Mancos pair, for they are almost the same color and texture, and doubtless represent a repeat run of marine and then continental conditions. On our map they are included in the Mesaverde Group.

At the town of Chimney Rock the Mesaverde Group is tilted steeply into a hogback. Pictured Cliffs Sandstone farther south is less steep, and forms a cuesta. This is the typical picture, for the strata level out southward, away from the San Juan uplift. Above the Pictured Cliffs Sandstone are several hundred feet more of Cretaceous lagoon and river sandstone and shale that signal the final eastward retreat of the sea and the beginning of uplift of the Southern Rockies. And above them lie Tertiary non-marine sandstone and shale derived from the San Juan Mountain uplift. The highway wanders back and forth through the different layers here, and because rock types and colors are often repeated it is hard to tell just where you are in the rock sequence. Some of the Tertiary rocks can be identified by their purple, green, and mustardy colors. Many contain volcanic ash mixed with pebbles and cobbles of volcanic rock. Cross bedding of the type that signifies shifting stream channels shows up well in some road-cuts.

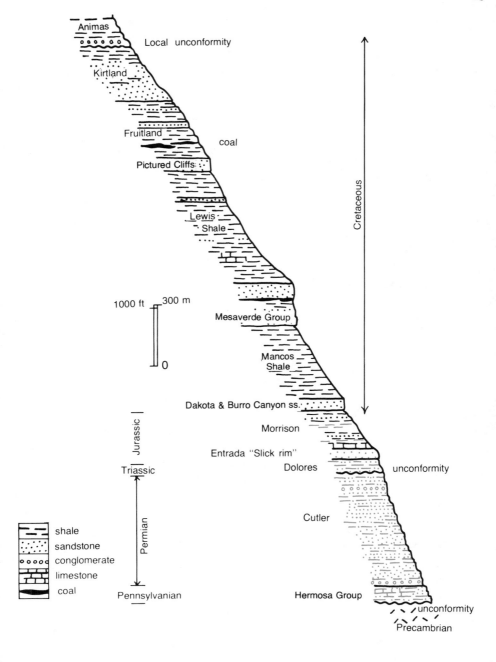

Stratigraphic column showing Paleozoic and Mesozoic rocks along U.S. 160 Pagosa Springs to Four Corners.

Durango lies between a cuesta of Mesaverde sandstone and the Dakota Hogback, in a wide spot where the Animas River crosses valley-forming Mancos Shale. Farther up the river older rocks, particularly red and gray sandstone and shale of Pennsylvanian, Permian, and Triassic age, rise toward the San Juan Mountains.

Durango was once a smelter town handling large shipments of ores from Silverton and other mining centers. It is the southern terminus of a narrow-gauge railway up the Animas River to Silverton. The railroad route passes through tilted sedimentary rocks that edge the San Juans, into the mountain core, where tiny mines cling precariously to awesome cliffs of Precambrian gneiss and schist.

u.s. 160
durango — four corners
(89 miles)

West of the Animas River at Durango, U.S. 160 passes north of some high-piled silver and uranium mill tailings and then goes up a canyon between hills and buttes capped with the lowest of the Mesaverde Group sandstones. This rock was laid down on the beaches and shores and bars of a retreating Cretaceous sea, not long before the Rockies began to rise. Coal occurs in the Mesaverde in minor quantity here but in thick seams farther north. In the middle of Mesaverde time, swamps and marshy lagoons edged many of the shallow shores.

North of the highway the sedimentary rocks tilt upward toward the San Juan Mountains. Occasional glimpses show the La Plata Mountains, a subsidiary range that contains a variety of intrusive rocks squeezed between the layers of Pennsylvanian and Permian red strata as sills and laccoliths or cutting across them as dikes and stocks. Individual bodies of igneous rock are small in these mountains, though some sills are several miles long. The La Plata Mountains may be the lower part of clustered volcanic vents that served as sources of some of the volcanic rock of the San Juans.

Landslides are common in the area west of Durango, where blocky, well jointed Mesaverde sandstone beds lie on top of weak and easily eroded, slippery-when-wet Mancos Shale. The Mancos is a marine

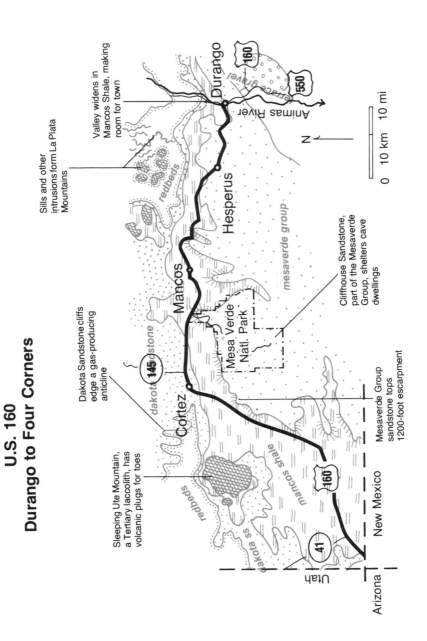

U.S. 160
Durango to Four Corners

Sills and other intrusions form La Plata Mountains

Valley widens in Mancos Shale, making room for town

Durango

160

550

Animas River

N

0   10 km   10 mi

redbeds

Hesperus

mesaverde group

Cliffhouse Sandstone, part of the Mesaverde Group, shelters cave dwellings

Mancos

Mesa Verde Natl. Park

Dakota Sandstone cliffs edge a gas-producing anticline

dakota sandstone

145

Cortez

Mesaverde Group sandstone tops 1200-foot escarpment

Sleeping Ute Mountain, a Tertiary laccolith, has volcanic plugs for toes

redbeds

mancos shale

dakota ss

160

New Mexico

41

Utah

Arizona

shale, and contains fossils of pelecypods, ammonites, and other sea-dwelling organisms. Here it is often silvery because it contains abundant tiny mica flakes.

The Mesaverde sandstones thicken westward, so the scarp at the top of the cuestas and mesas south of the road gets higher and higher in that direction. Cliff houses of Mesa Verde National Park are in the Cliffhouse Sandstone, the uppermost third of the Mesaverde Group (see MESA VERDE NATIONAL PARK). Lower mesa slopes are composed of Mancos Shale, here about 800 to 1000 feet thick. The highway passes through the town of Mancos, namesake of the dark gray shale. Mancos is one of the oldest coal mining towns of western Colorado, and now is an agricultural community as well; the fine, sometimes sandy shales form good soil.

A few miles west of Mancos, on the summit of a gentle anticline near the entrance to Mesa Verde National Park, a small gas field produces from shallow wells drilled into the Dakota Sandstone, below the Mancos Shale. A number of other anticlines in the area west and northwest of Mesa Verde produce gas, sometimes just enough for local use. Some produce carbon dioxide, which is made into dry ice or used to build up pressure in oil wells. Like oil, gas tends to rise through porous rock until it meets with some impenetrable shale. Anticlines form a natural trap for an upside-down gas pool.

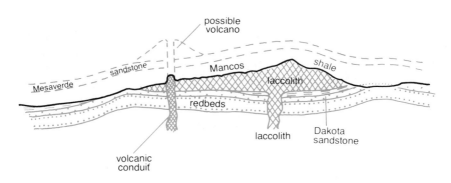

North-south section through Sleeping Ute Mountain showing laccolith intruding Mancos Shale.

Cortez lies on flat layers of Dakota Sandstone that no longer reflect the San Juan uplift. Beyond Cortez to the southwest is a picturesque landmark, named by Indian lore Sleeping Ute Mountain. This little range is a Tertiary laccolith, with a pair of resistant volcanic necks, once the conduits of volcanoes, forming the Ute's toes. The igneous rock domes up layers of Mancos Shale and Mesaverde Group sandstone. Underlying Dakota Sandstone and redbeds don't bend up around the base of the laccolith, so they must continue beneath it to form its floor.

From Cortez the highway heads south across the Ute Indian Reservation between Mesa Verde and Sleeping Ute Mountain, descending gradually through the lower half of the Mancos Shale. In the distance to the north the symmetrical tip of Lone Cone marks the west end of the San Juan Mountains. Shiprock, in New Mexico, one of the most picturesque and famous of volcanic necks, can be seen in the distance almost straight ahead. Mountains west of it are on the New Mexico-Arizona line.

An isolated pinnacle of Mesa Verde sandstone stands like a sentinel near the southern end of Mesa Verde. Once it was part of the main mesa mass. Though wind is an important erosional agent in this desert land, running water sculptured the corrugated shale slopes.

R.W. BROWN PHOTO, COURTESY OF USGS

This part of the route offers vistas into the plateau country where four states meet — Arizona, Colorado, New Mexico, and Utah. In this desert area wind takes a strong part in shaping the land. Dust- and sandstorms carry silt particles and sweep sand grains before them, using them to abrade rock surfaces. Often the wind cleans away all the fine particles, leaving only a rock-strewn surface or **desert pavement**. Elsewhere it deposits dunes, blows out hollows in sandy soil, or carves fist-sized holes in sandstone.

Because of the lack of moisture, the rate of erosion here is usually slow. But in soil not protected by vegetation, sudden floods wreak tremendous changes in a few short hours — though perhaps only once in many years. Heat and cold play a part, for desert nights are notoriously chilly, often cold enough to fracture rock by volume changes that accompany sudden cooling. Desert rocks develop **desert varnish**, dark glossy coatings of manganese and iron drawn to the surface by heat and occasional moisture. Watch for these features as you approach Four Corners.

The spot where four states meet (the only such spot in United States) is on a small plateau of Dakota Sandstone, surrounded by the interesting scenery of the plateau deserts. At Mile 4, U.S. 160 starts its descent from the Mancos Shale through older Cretaceous rocks, including a white band of limestone and the blocky Dakota Sandstone in channels and gullies.

mesa verde national park
(48-mile side trip from u.s. 160)

Mesa Verde National Park was established to preserve and display unusual archeological ruins: cave dwellings, rock paintings, and surface sites of early occupants of the area. However the site itself — the great gray-sloped mesa with its virtually impregnable fortress of Mesaverde Sandstone — is as distinctive geologically as the cave-sheltered apartment houses of the ancient Indians are archeologically.

The Mancos Shale, forming the lower slopes of Mesa Verde, is well exposed along the entrance road. Brownish gray shale with thin sandstone beds (sometimes offset a foot or two by small faults) slides easily, as you can tell from slumps right along the road. It is a marine shale and contains occasional fossil clams and ammonites. It formed at a time when the sea spread across nearly all of Colorado. Above it, deposited as the sea retreated, are shoreline sandstones of the Mesaverde Group.

The Mesaverde Group here consists of three formations, the basal shoreline or shallow-sea Point Lookout Sandstone; the shale-coal sequence of the Menefee Formation deposited in lagoons, marshes, and swamps close to sea level; and the upper Cliffhouse Sandstone,

296

Slope-forming Mancos Shale and cliff-forming sandstones of the Mesaverde group characterize Mesa Verde National Park. Erosion of the soft shale undermines the more resistant, blocky sandstone layers, forming cliffs. L.C. HUFF PHOTO, COURTESY OF USGS

another shoreline marine deposit. The middle shale-coal part, softer and weaker than rocks above and below it, erodes easily and often undermines the massive, light-colored Cliffhouse Sandstone. Weakened, the sandstone falls off in blocks or spalls off in great arcs to make the arched caves that characterize most of the cliff-dwelling sites. Since the rock weathers faster when it is wet, many of the caves have formed near springs where rainwater, percolating from above through the porous sandstone, reaches the less permeable shales and starts to flow sideways along the layers. What could be more convenient for a cave dweller? The massive lower sandstone sufficed to keep away enemies and provided a ready disposal system — refuse was just tossed out, a convenience also for the archeologist searching for clues to the daily life of the early Americans.

Coaly shales and tan sandstones above the two massive cliffs are higher parts of the Mesaverde Group. These are exposed along the entrance road but have been stripped off by erosion near the cliff dwellings.

The drainage pattern of streams draining Mesa Verde is typical of plateau areas. Each major stream branches again and again to form a tree-like or dendritic pattern, as shown on the relief map in the Visitor Center. Erosion is not severe here now, but during times of more intense rainfall, as in the rainy cycles that accompanied Ice-

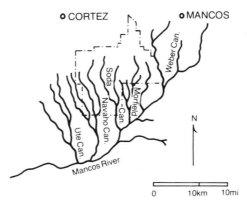

Dendritic drainage of Mesa Verde. Because the streams follow joints in the rock, many are parallel.

Age glaciation, each small stream worked its way headward into the plateau, branching and rebranching often, following joints in the rock. In the process, caves formed (and later collapsed) as spring-fed streams undermined the massive caprock. By the time of man's arrival here, the climate had become drier and the caves were in less danger of collapsing. The first-comers lived on the surface of the plateau, building pit houses and farming or hunting on the mesa surface. Much later, the cave dwellings were built, perhaps as protection from both weather and enemies. Cave dwellings were occupied less than a hundred years before prolonged drought in the 13th century caused their abandonment.

Cliff Palace at Mesa Verde was built on the sloping floor of a deep cave in the Cliffhouse Sandstone. Like other large once-inhabited caves in the National Park, this one formed where springs emerging at the top of the Menefee Formation weakened and undermined the massive sandstone above.

M.A. PISCHEL PHOTO, COURTESY OF USGS

298

Views of the Four Corners plateau country to the south and west are exceptionally good from a number of points along the entrance roads between the ruins. Counterclockwise from Sleeping Ute Mountain to the west are:

• The Lukachukai and Chuska Mountains along the New Mexico-Arizona border.

• Shiprock, a large volcanic neck in northwestern New Mexico, often obscured by smog from the Four Corners power plant. (Unfortunately, the plant is a product of geology as well; it burns Cretaceous coal from the Mesaverde Group.)

• The fingers of Mesa Verde to the south, pointing into the San Juan Basin.

• The Nacimiento Mountains of New Mexico blending northward with the southern San Juans in Colorado.

• The La Plata Mountains to the northeast and Rico Mountains to the north. In the distance northward, Lone Cone marks the west end of the San Miguels, part of the San Juan Range.

• The Abajo Mountains of Utah, in the distance north of Sleeping Ute Mountain.

co. 13/789

craig — rifle

(90 miles)

As Colorado 13/789 leaves U.S. 40 just west of Craig, it heads south toward the White River Plateau, one of the three prominent volcanic areas west of the long ranges of the Colorado Rockies. For some distance it travels on tan sandstone and gray shale of the Mesaverde Group, which is usually divided into three formations, a lower sandstone unit, a middle shale and coal unit, and an upper unit that is predominantly sandstone. Seven miles south of Craig, Mesaverde Group coal is being mined on the northwest slope of Breeze Anticline (see U.S. 40 STEAMBOAT SPRINGS - CRAIG) to fuel a nearby power plant. Power lines from this plant parallel CO 13/789 all the way to Rifle.

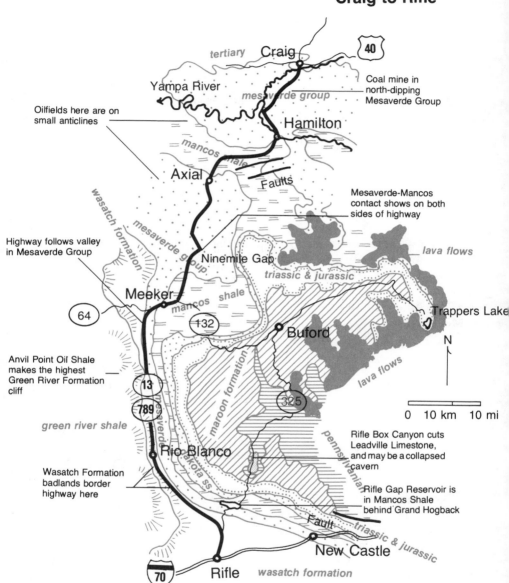

Colorado 13/789
Craig to Rifle

tertiary

Craig

40

Coal mine in
north-dipping
Mesaverde Group

Yampa River

mesaverde group

Oilfields here are on
small anticlines

Hamilton

mancos shale

Axial

Faults

Mesaverde-Mancos
contact shows on both
sides of highway

lava flows

Highway follows valley
in Mesaverde Group

wasatch formation

mesaverde group

Ninemile Gap

triassic & jurassic

Meeker

mancos shale

64

132

Trappers Lake

Buford

N

Anvil Point Oil Shale
makes the highest
Green River Formation
cliff

maroon formation

lava flows

13

325

789

0   10 km   10 mi

green river shale

mesaverde group

pennsylvanian

Rifle Box Canyon cuts
Leadville Limestone,
and may be a collapsed
cavern

Rio Blanco

dakota ss.

Wasatch Formation
badlands border
highway here

Rifle Gap Reservoir is
in Mancos Shale
behind Grand Hogback

Fault

triassic & jurassic

New Castle

70

Rifle

wasatch formation

There are other coal mines along the route, all recovering coal from the middle part of the Mesaverde Group. Coal seams show well in both old and new highway cuts, particularly at Mile 77.

At Hamilton, the route crosses into an older though still Cretaceous rock, the Mancos Shale. Mesaverde sandstones top the hills near the highway.

Swinging west, the route rounds two small but interesting domes where Mancos rocks are bent up and then eroded partly away. Watch for steep dips in a resistant sandstone layer of the Mancos Shale, and for oil wells in the valleys eroded in the centers of the domes. Because shales erode more easily than sandstones, and because the crests of these structural domes have been eroded off, ring-shaped ridges of sandstone still stick up like super-sized broken eggshells around the domes.

Near Axial the highway gets into Mesaverde Group sandstones again, with their characteristic coal beds. Another typical feature of the Mesaverde Group is the way in which the sandstone cliffs and ledges develop numerous little rounded recesses, like so many swallows' nests. Grains of sand, initially loosened by normal weathering processes associated with rain and other moisture, blow away on windy days, leaving small hollows. Partly because they hold moisture that breaks down the minerals that cement the sand grains together, and partly because grains of sand bombard their neighbors as they are whirled by wind eddies in the hollows, the holes deepen, sometimes until they are deeper than they are wide. This type of weathering is called **honeycomb weathering,**

Burning coal seams ignited by natural causes have in some places baked the layers of shale, turning them a bright brick red like fired pottery. Watch for small anticlines and synclines exposed along the highway.

East of the road, rocks bend up in a large monocline, a simple fold that edges the White River Plateau. About 50 miles across, this plateau seems to be transitional between the linear faulted anticline ranges farther east and the true plateaus farther west. Partly surfaced with Pennsylvanian and Permian sedimentary rocks (younger rocks having been erodeed off), and partly topped with Tertiary lava flows, its scenery is markedly different than that of other Colorado mountains.

The cliffs west of Meeker are the north end of the Grand Hogback, where Mesaverde Group sandstone and shale turn up steeply along

Colorado 132, branching from 13/789 near Meeker, makes an interesting 55-mile side trip. It follows the north fork of the White River through the Dakota Hogback and successively older sedimentary rocks, and finally climbs into volcanic rocks near Trappers Lake, at the edge of Flattops Wilderness Area. JACK RATHBONE PHOTO

the monocline that borders the White River Plateau. The hogback extends south in a big S-shaped curve as far as Redstone, 30 miles south of Glenwood Springs. Colorado 13/789 goes through the Grand Hogback west of Meeker and follows it almost to Rifle, running first in a valley eroded in a shaly part of the Mesaverde Group and then in soft shales at the base of the Wasatch Formation. The Mesaverde Group along Grand Hogback is about 6000 feet thick, and represents rapid, abundant sedimentation along a sandy coast, with numerous marshes and swamps where plant material — future coal — accumulated.

South of about Mile 30 the highway works its way across the uppermost Mesaverde beds and the lowermost conglomerate and sandstone beds of the Wasatch Formation, which is early Tertiary (Paleocene) in age. High cliffs west of the road are formed of flat-lying grayish pink layers of this formation., and of overlying gray or yellow-gray Green River Shale (Eocene). Near the top of the cliff is one of the world's largest untapped energy reserves, the Green River oil shale (see I-70 RIFLE — UTAH LINE). The most promising oil shale areas are underground in untraveled parts of the Roan Plateau west of Rio Blanco.

Between Miles 15 and 16 white sandstone at the base of the Wasatch Formation is eroded into badlands and fantastic tepee-shaped mounds. Exposures of the Green River-Wasatch cliff improve southward as vegetation decreases. Sheer cliffs near the top of the plateau contain the rich oil-shale beds, whose browner hue can be detected even from this distance.

From this area also, the Elk Mountains can be seen to the south, as can Grand Hogback arcing southeastward beyond the Colorado River. Notice the sloping pediments that surround Battlement Mesa, a Wasatch-Green River Formation mesa capped with Pliocene basalt flows. The words pediment and piedmont both derive from the same Latin roots; both mean "foot of the mountains." But pediment has been given a special meaning by geologists: an eroded surface at the foot of a mountain, carved right into the mass of the mountain itself.

Government Creek parallel to the highway follows the curve of Grand Hogback, seeking out the easiest route to the Colorado River. Its course is quite obviously controlled by geologic structure, as is the course of Rifle Creek on the other side of Grand Hogback.

For a close view of Grand Hogback and the south side of the White River Plateau, take a 26-mile round trip on Colorado 325 to Rifle Gap, Rifle Falls, and Rifle Creek Box Canyon. Rifle Gap was the site of a well publicized but ill-fated orange nylon "sculpture" a few years ago. In it, exposures of the Mesaverde Group are unsurpassed. The contact with the overlying Wasatch Formation is clearly exposed, and every resistant sandstone bed stands out in sharp relief. With the help of a small dam, Grand Hogback holds back the waters of Rifle Gap Reservoir, lying in a valley of Mancos Shale. Crossing this valley the road goes through the Dakota Hogback, purple and green shales of the Jurassic Morrison Formation, some Triassic rocks, and a full array of Paleozoic sediments. For a diagram of the rocks here, see I-70 DOTSERO — RIFLE.

At Rifle Falls, the creek cascades over a travertine dam made of calcium carbonate precipitated by the stream as it emerged, laden with chemicals, from the limestones of Rifle Box Canyon. Narrow, high-walled parts of the canyon, possibly once a long subterranean cavern, are overhung by cliffs of Mississippian Leadville Limestone. In places the cliffs and caves contain lush oases of shade-loving plants and offer welcome relief on hot summer days.

Colorado 141
Whitewater to Naturita

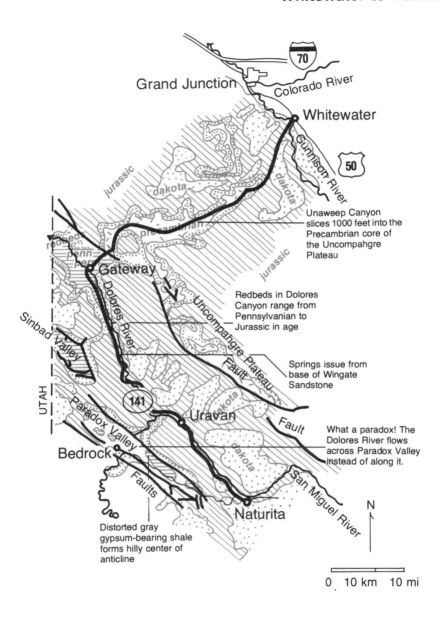

Grand Junction

Colorado River

Whitewater

70

50

Gunnison River

jurassic

dakota

dakota

precambrian

jurassic

redbed

penn
perm

Gateway

Dolores River

Sinbad Valley

UTAH

Paradox Valley

141

Uravan

Uncompahgre Plateau Fault

dakota

Bedrock

Faults

San Miguel River

Fault

Naturita

Unaweep Canyon slices 1000 feet into the Precambrian core of the Uncompahgre Plateau

Redbeds in Dolores Canyon range from Pennsylvanian to Jurassic in age

Springs issue from base of Wingate Sandstone

What a paradox! The Dolores River flows across Paradox Valley instead of along it.

Distorted gray gypsum-bearing shale forms hilly center of anticline

N

0   10 km   10 mi

co. 141
whitewater — naturita
(100 miles)

Leaving Whitewater and crossing the Gunnison River, Colorado 141 plunges almost immediately into Unaweep Canyon, one of the most unusual and interesting canyons in Colorado, and passes crosswise through the very heart of the Uncompahgre Plateau. The plateau is a twice-uplifted block of Precambrian rocks overlain by Mesozoic sedimentary layers. First lifted by faulting in Pennsylvanian time, this area was part of Uncompahgria, one of two island ranges of the Ancestral Rockies. The present southwest side of the uplift approximately coincides with that of the ancestral highland; however Uncompahgria extended much farther east and north than its modern counterpart. It was eroded down to its roots in Permian and Triassic time, and furnished sand and gravel and mud to the redrock formations of southwest Colorado. These red sandstones and shales lie now right on Precambrian granite, and are overlain by horizontal layers of Jurassic and Cretaceous sedimentary rocks described under I-70 RIFLE — UTAH LINE.

After deposition of these rocks the plateau was not lifted again for some time, probably not until late in Tertiary time, during Miocene-Pliocene regional uplift 28 to 10 million years ago.

Colorado 141 follows a deep slot in the plateau called Unaweep Canyon, now occupied by small streams of East Creek, flowing east, and West Creek, flowing west. Unaweep Canyon slashes like a knife through a layer cake, down to Triassic redbeds and eventually deep into the Precambrian faulted core of the uplift. As you will see, the western part of the canyon cuts 1000 feet or more into the granite and metamorphic "basement" rocks.

A canyon of this depth is certainly out of proportion to the little streams that wind along its floor today. What then could have carved it? The little streams with more water? Apparently not, for they head in the flat bottom of the canyon, separated by an almost imperceptible divide near Mile 135. The only rivers around that are capable of carving such a deep canyon are the Colorado and the Gunnison, both

Unaweep Canyon, a deep slot cutting southwest directly across the Uncompahgre Plateau, holds secrets not yet fully deciphered. Walled with hard Precambrian granite and metamorphic rocks of the core of the plateau, it must have been cut by a river far more powerful than the two minuscule streams that drain it today.
S.W. LOHMAN PHOTO, COURTESY OF USGS

of which, coming together at Grand Junction, now flow around the north end of the Uncompahgre Plateau.

Geologists who have studied the area feel this canyon must be the work of one or both of these rivers. Rushing along a course established across Tertiary sediments that once blanketed the plateau completely, the river or rivers entrenched a channel through the soft Tertiary sediments and through harder Mesozoic rock layers, and finally into the still harder Precambrian rocks.

Why the rivers eventually abandoned this route can also be guessed at. The present Uncompahgre Plateau became a distinct feature late in Tertiary time, long after the river course was established and the canyon cut. Uplift must have been so rapid at that time as to make it difficult for the river to maintain its course. Some small stream east of the uplift, flowing north and eroding easily and rapidly through soft Tertiary and Cretaceous shales, "pirated" the larger stream and "captured" its flow.

Unaweep Canyon seems to line up with the course of the Colorado River above Palisade, so many authors think the Colorado is responsible for the canyon. However old river-deposited gravels in Unaweep Canyon consist mostly of volcanic and sedimentary rocks like those of the Gunnison River today. It may be that the Colorado carved the canyon initially, with the Gunnison as a tributary. At Cactus Park another deep canyon comes in from the east — is this an ancestral Gunnison canyon?

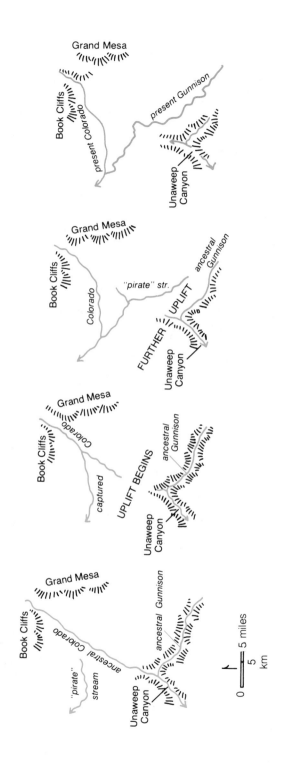

Four stages in the life of Unaweep Canyon – one interpretation

307

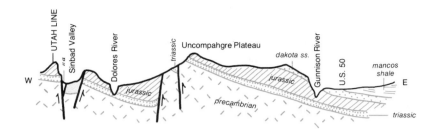

Section across Uncompahgre Plateau just south of Colorado 141 and Unaweep Canyon

Sedimentary rock layers in Unaweep Canyon and in the area southwest of the Uncompahgre Plateau are (from top to bottom):

• Dakota Sandstone (Cretaceous), about 130 feet of light brown sandstone visible at Mile 152.

• Morrison Formation (Jurassic), 50-100 feet of red and green and sometimes bluegreen shale and sandstone.

• Summerville Formation (Jurassic), 50 feet of red and green mudstone and siltstone, often considered part of the Morrison. This formation contains uranium and vanadium.

• Entrada Sandstone (Jurassic), 50 feet of crossbedded sandstone forming a massive peach-colored or tan cliff known locally as the "slick rim." Its base is irregular, sometimes filling channels in underlying beds. This formation is probably wind-deposited, with sweeping dune-type crossbedding.

• Kayenta Formation (Jurassic), 100 feet of hard river-deposited sandstone with red or purple siltstone streaks, forming a series of ledges.

• Wingate Sandstone (Triassic), 350 feet of wind-deposited sandstone forming great vertical red cliffs (Mile 143 top).

• Redbeds (Permian and Triassic), bright red to purple sandstone and shale visible for many miles in Dolores River Canyon. The term "redbeds" is a favorite in southwestern United States, perhaps because it covers a multitude of sins. It is usually applied to red and pink sandstone and shale, sometimes with thin limestone layers, nearly always with some gypsum, of late Pennsylvanian to Jurassic age — sediments in Colorado quite obviously derived from the Ancestral Rockies. These rocks are usually thought of as being floodplain deposits similar to the widespread deposits of large rivers like the Mississippi and Amazon today. Their redness is due to fine particles of iron oxide, particularly the mineral hematite, which coats sand grains and mingles with silt and clay grains in the rock.

At Mile 143-142 the canyon cuts into Precambrian granite and metamorphic rocks. Near Mile 130 notice the banding on these rocks, a banding that strongly suggests shale and sandstone layers from which the rock must have been derived. These rocks are also strongly veined with light-colored pegmatite dikes; often they are jointed as well. They rise westward into higher and higher cliffs until at the west end of Unaweep Canyon they are cut off abruptly by the large fault that edges the southwest side of the Uncompahgre Plateau. Triassic rocks of the Wingate Formation show in cliffs to the south, and must bump right into the Precambrian rocks, but the details of the fault are hidden by broken rock debris, soil, and vegetation.

From Gateway to Naturita the highway goes up the Dolores River, threading its way through a narrow canyon in the same colorful rock layers. Because of the desert climate the geology is extremely well exposed. The sedimentary rocks are essentially horizontal as in most of the plateau country. West into Utah, and southwest to Grand Canyon and beyond, horizontal strata predominate, and the occasional faults and very simple folds take second place as landscape

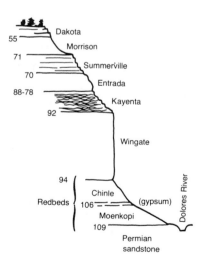

Dolores River Canyon. Figures give highway miles where contacts are at road level.

On the narrow ridge-like butte called the Palisade, north of Gateway, the sedimentary rock sequence extends from the Morrison knob at the north end down to Triassic redbeds and even, between the town and the banks of the Dolores River, some Permian redbeds. The highest cliffs on the Palisade are Wingate Sandstone.

F.W. CATER PHOTO, COURTESY OF USGS

makers. In this colorful land of lonesome beauty, most of the rocks are Mesozoic and Permian. They vary greatly in color, but the warm tones — pink, peach, red, and a deep rich reddish purple — always dominate.

As you go south up the Dolores River you climb gradually through the same strata listed above. The accompanying sketch shows the miles at which each contact is at roadside level.

The Morrison, Summerville, and Chinle Formations contain quite a lot of uranium, usually as very fine yellow particles of carnotite that coat individual sand grains and fill spaces between them. Frequently the uranium adheres to sand grains and bits of wood that were deposited in stream channels. Such ore bodies are long and slim, and can be discovered and followed only through careful geologic mapping. The town of Uravan takes its name from two mineral products that brought it into existence, uranium and vanadium.

From Vancorum, Colorado 90 to the right (west) will take you on an interesting side trip of just a few miles into aptly named Paradox Valley, an unusual collapsed salt anticline. Pennsylvanian rocks of southwestern Colorado contain salt and gypsum and potash layers formed by evaporation of seawater in a land-locked basin. The evaporites are thickest along the southwest side of the ancestral island range of Uncompahgria; 5000 feet or more were deposited here during well documented cycles when the basin was cut off from the sea.

310

Both salt and gypsum flow in a solid state like silly putty, but movement is very very slow. Weighed down by overlying rocks they try to escape upward, pushing against overlying layers and arching them into salt anticlines. Three such anticlines developed in this area: Paradox Valley, Sinbad Valley, and Gypsum Valley. All are long folds with a northwest-southeast trend that has been determined by the position of faults in Precambrian rocks.

Sometime after the salt anticlines formed, faults developed along their margins. For as the salt was dissolved away by groundwater the anticlines collapsed, leaving broken Cretaceous rocks and crumpled gray gypsum-bearing Pennsylvanian sedimentary layers in the centers of deep valleys edged with cliffs of Wingate Sandstone.

Near Uravan, the rounded "slick rim" Entrada Sandstone lies on a flat-bedded, blocky formation called the Navajo Sandstone. Both are Jurassic in age. The Navajo extends west from here all the way to Zion National Park in Utah. F.W CATER PHOTO, COURTESY OF USGS

311

Colorado 145
Naturita to Telluride

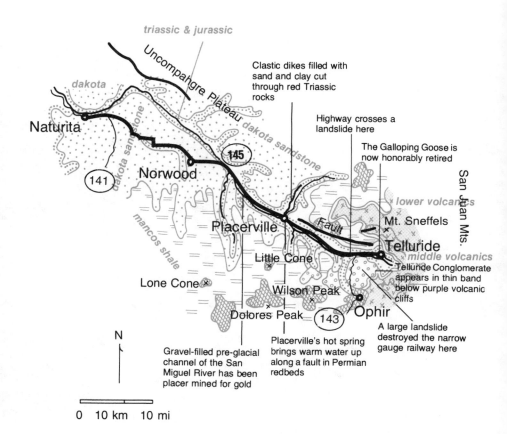

triassic & jurassic

dakota

Uncompahgre Plateau

Clastic dikes filled with
sand and clay cut
through red Triassic
rocks

Highway crosses a
landslide here

The Galloping Goose is
now honorably retired

Naturita

dakota sandstone

145

141

Norwood

mancos shale

lower volcanics

San Juan Mts.

Placerville

Fault

Mt. Sneffels
×

Telluride

middle volcanics

Little Cone

Telluride Conglomerate
appears in thin band
below purple volcanic
cliffs

Lone Cone ×

Wilson Peak
×

Dolores Peak
×

143

Ophir

A large landslide
destroyed the narrow
gauge railway here

N

Gravel-filled pre-glacial
channel of the San
Miguel River has been
placer mined for gold

Placerville's hot spring
brings warm water up
along a fault in Permian
redbeds

0   10 km   10 mi

co. 145
naturita — telluride
(54 miles)

Naturita lies at the southeast end of the Paradox Valley salt structure described in the preceding section. Look back along that valley from the top of the hill a mile from Naturita to see Mesozoic rocks tilted by the upward flow of salt.

The general trend of both topography and structure in this area is northwest-southeast. To the north the Uncompahgre Plateau parallels this trend. It is surfaced with Dakota Sandstone faulted several thousand feet above the corresponding Dakota surface of the Naturita area.

Between Naturita and Norwood the highway crosses this Dakota surface, shortcutting a bend in the San Miguel River. Ahead you can see Mt. Sneffels, Wilson Peak, and Dolores Peak in the San Juan Mountains, as well as Lone Cone a little farther south. All the highest peaks of this part of the San Juans are made of intrusive rock, possibly formed in conduits that once led upward to volcanoes.

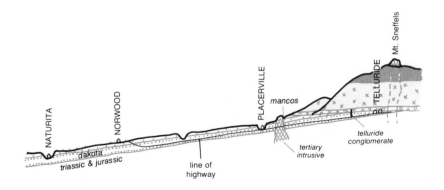

Section along Colorado 145 Naturita to Telluride

At Norwood the highway drops into a canyon carved by the San Miguel River in rocks of Triassic and Jurassic age. The sedimentary formations exposed along this road are almost identical with those described in the preceding log, but they are more often cloaked in vegetation as you approach the San Juan Mountains, which form a sort of storm center where rainfall is more abundant. At Mile 87 the redbeds at the base of the sequence show up well across the river, where they include a purple conglomerate packed with pebbles of almost uniform size.

A number of almost vertical faults cut the sedimentary rocks here. At Mile 87, for instance, one runs along the north wall of the valley. The north side is downdropped as much as 500 feet, and the valley is lopsided as a result! Another at Mile 90 cuts across the canyon. It has a vertical displacement of about 200 feet, bringing Morrison Formation against lower redbeds.

At Mile 89 a boulder-filled channel occupied by the San Miguel River in pre-Ice-Age times can be seen up and across the river. Boulder heaps below it are from placer workings that processed the old gravels for gold.

Between Miles 78 and 77 a large dike cuts obliquely across the canyon. It is well exposed near the road. The flat face of the dike shows up in a side gull just beyond the spot where the dike crosses the highway. Dikes become more and more frequent approaching the San Juans, reminders of the volcanic nature of those mountains. Some dikes are hard and weather as ridges. Others, softer and more easily eroded, weather into slots between walls of baked sedimentary rocks.

Small mines east of Placerville are vanadium mines, some dating from as long ago as 1910 when vanadium was used only to color red and orange china glazes and glass. Although small amounts of vanadium are widely distributed in various minerals as well as in coal and petroleum, vanadium ores of commercial value are rare. The mineral is now used in high-strength high-performance steel for power tools and construction.

East of Placerville watch for moraines of the San Miguel glacier and of tributary glaciers coming in from south and north. As you will see, the San Miguel Glacier headed up this valley above Telluride. The lowest moraine, an irregular hill cutting across the valley floor, is at Mile 74 at an elevation of about 8500 feet. The valley is distinctly U-shaped from about Mile 76 — a sure sign of glacial erosion. For more on glacial features see Chapter IV.

Landslides are common in this region, with four factors operating in their favor:

- Unstable volcanic rocks often contain thick layers of poorly consolidated volcanic ash.

- These volcanic rocks lie above even less stable Mancos Shale.

- Valley walls were over-steepened by glaciers.

- In this area mountain storms and heavy snows often saturate the soil and underlying rocks, adding weight and slipperiness.

Of course man, with his roadcuts and railroad cuts and mines, has contributed his share to landslide formation too. The highway crosses a slide near Society Turn, the junction to which Telluride high society used to ride their carriages on Sunday afternoons. Other slides visible across the valley terminated the career of the "Galloping Goose," the contraption (no other word fits) that used to ply the narrow-gauge railroad line to Ridgeway.

The Telluride spur of Colorado 145 continues up the glaciated valley of the San Miguel, its floor filled with more than 500 feet of Pleistocene lake deposits formed at a time when the river was dammed by glacial moraines. U-shaped hanging valleys and well developed glacial cirques frame the town of Telluride. The hanging valleys are those of small glaciers that were unable to keep up, in terms of erosive power, with the bigger glacier that occupied the main valley of the San Miguel. Bridal Veil Falls cascades from one of them, with a free fall of 350 feet.

Mesozoic rocks rise eastward toward the San Juans, where they were lifted into a broad dome sometime late in the Cretaceous Period, in the first throes of the Laramide Orogeny. The dome was later beveled by erosion, so rocks above the Dakota Sandstone, though present above Society Turn, do not appear at the east end of the valley. The valley walls above Telluride clearly display the **angular unconformity** where the beveled surface of tilted Mesozoic rocks is overlain by horizontal Telluride Conglomerate, the latter about 250 feet thick and forming a reddish vertical cliff. Above is the San Juan Tuff, part of what we are calling the first-phase volcanic rocks of the San Juans (see Chapter V). This tuff is a deposit of volcanic ash and cinder 1200 to 1500 feet thick; it covers an area more than 100 miles across.

During the second volcanic phase, thousands of feet of lava flows and ash falls were superimposed on the San Juan Tuff. The long ridge on the skyline north of Telluride, and Greenback Mountain around the head of the valley to the south, are capped by second-phase volcanic rocks.

Unfortunately, mill tailings disfigure the beautiful glaciated valley above Telluride. Idarado Mine, until recently the largest active mine in this area, produced 1650 tons of gold, silver, lead, copper, and zinc ores **daily**. In its heyday, Telluride was the site of literally hundreds of miles of mine tunnels, and hundreds of mine portals, many of them now concealed by second-growth spruce and aspen forest. (The original forest was cut for mine timbers, buildings, and fuel.) The mines produced gold, silver, lead, and zinc, all in veins or dikes associated with the many igneous intrusions in this region. One mine penetrated all the way to Red Mountain, on the other side of the mountain about seven miles away. Many of the old veins have until recently been worked through cross-cut tunnels by the Idarado Mine.

Glossary

Alluvial fan: a cone-shaped mass of gravel and sand deposited by a mountain stream where it runs out onto a level or nearly level plain.

Alluvium: sediments formed by rivers and streams.

Ammonites: an extinct group of shell-forming mollusks related to the living Chambered Nautilus. Both coiled and straight-shelled forms are known.

Andesite: a medium-colored volcanic rock containing a high proportion of feldspar.

Anticline: a fold that is convex upward. When eroded, an anticline has the oldest rocks in the center.

Aquifer: a porous rock layer from which water may be obtained.

Artesian well: a well in which water level rises above the top of the water-bearing layer.

Basalt: a dark-colored volcanic rock that often contains small round vesicles or gas bubbles.

Basement: igneous and metamorphic rocks, usually of Precambrian age, lying below the sedimentary rock sequence.

Batholith: a very large mass of igneous rock (larger than 40 square miles), intruded as molten magma, often formed at least in part by melting and recrystallization of older rocks.

Bedrock: solid rock exposed at or near the surface.

Bentonite: clay formed from decomposition of volcanic ash.

Beryl: a light green mineral occurring in pegmatites; emerald and aquamarine are gem varieties.

Biotite: black (iron-rich) mica.

Blowout: a cup-shaped hollow in sand or silt formed by wind erosion, often many feet in diameter.

Breccia: volcanic rock consisting of fragments ejected from a volcano, lying in a fine matrix of volcanic ash.

Butte: an isolated hill or small mountain, often with a horizontal top and steep sides.

Calcite: a mineral, calcium carbonate ($CaCO_3$), the principal constituent of limestone.

Caldera: a large basin-shaped volcanic depression formed by explosion or collapse of a volcano.

Caprock: a comparatively resistant rock layer — either sedimentary or volcanic — forming the tops of mesas, buttes, and cuestas.

Carbonaceous: containing carbon or coal derived from organic material.

Carbonate: rocks containing carbon and oxygen in combination with sodium, calcium, or other elements, particularly as in limestone or dolomite.

Cinder cone: a small conical volcano formed by accumulation of volcanic ash and fine cinder around a volcanic vent.

Cirque: a deep, steep-walled, usually semicircular scoop in a mountain excavated by the head of a glacier.

Columnar jointing: vertically arranged, polygonal joints due to shrinkage accompanying cooling of lava and ash flows.

Concretion: a pebble-shaped or nodular concentration of minerals deposited around a central nucleus, usually harder than the surrounding rock.

Conglomerate: rock composed of rounded waterworn fragments of older rock, usually in combination with sand.

Cross-bedded: with laminae slanting obliquely between the main horizontal layers of a sedimentary rock (generally sandstone).

Cuesta: a hill with a long gentle slope formed by a resistant caprock, and a short steep slope on cut edges of underlying weaker rock.

Dacite: a volcanic rock with a high proportion of quartz and feldspar.

Dendritic drainage: a tree-like pattern of irregularly branching streams.

Desert Pavement: a surface veneer of pebbles and stones resulting when finer dust and sand are blown away by wind.

Desert varnish: a dark shiny surface of manganese and iron oxides that characterizes many exposed rock surfaces in deserts.

Dike: a thin body of igneous rock resulting when liquid magma intrudes a vertical joint in the rocks.

Dip: the angle at which a rock layer is inclined below the horizontal.

Dolomite: a limestone-like rock containing magnesium carbonate as well as calcium carbonate.

Earthflow: a slow flow of soil lubricated with water.

Escarpment: a cliff or steep slope edging a region of higher land.

Evaporite: a mineral deposited from highly mineralized or salty water as a result of evaporation.

Extrusive rocks: igneous rocks that cool on or very near the earth's surface; volcanic rocks.

Fault: a break in the rock along which rocks on either side have moved relative to each other.

Fault scarp: a cliff formed by a fault, usually modified by erosion.

Feldspar: a group of abundant light-colored rock-forming minerals.

Flatirons: triangular-shaped remains of hogback ridges steeply tilted against the flank of a mountain.

Formation: a named, recognizable, and mappable unit of rock.

Geomorphology: a branch of geology which deals with the earth's surface features or landforms.

Gneiss: a coarse-grained metamorphic rock with alternating bands of granular crystalline minerals such as quartz and feldspar, and fine, often platy dark minerals such as biotite.

Gouge: fine and often fairly soft clayey material along the wall of a vein or between the walls of faults.

Granite: coarse-grained intrusive igneous rock with feldspar and quartz as principal minerals.

Ground moraine: material deposited on the ground surface by a melting glacier.

Groundwater: subsurface water filling rock pore spaces, cracks, or solution channels.

Group: a stratigraphic unit consisting of several formations, usually originally a single formation subdivided by subsequent research.

Gypsum: a common evaporite mineral ($CaSO_4$), the main ingredient of plaster.

Hanging valley: a valley whose floor is substantially higher than the floor of the valley into which it leads.

Hematite: and ore of iron (Fe_2O_3).

Hogback: a sharp ridge produced by erosion of highly tilted rock layers, one of which is more resistant than the others.

Honeycomb weathering: weathering of rock (usually sandstone) by wind and water, producing deep fist-sized holes in the rock surface.

Hornblende gneiss: dark gneiss containing hornblende as the most abundant mineral.

Hydrostatic pressure: pressure caused by weight of water in water-bearing rock layers.

Hydrothermal: caused by hot water.

Icecap: glacial ice which spreads in all directions over a high, relatively flat surface.

Igneous rock: rock formed by solidification of molten rock material.

Incompetent: relatively weak and unable to support its own weight or the weight of overlying material.

Injection gneiss: gneiss containing sheets and veins of granite injected under great pressure deep in the earth's crust.

Intrusive rock: igneous rock that has hardened from molten rock material (magma) before reaching the surface.

Joint: a fracture in rock along which no appreciable movement has occured.

Kaolin: a type of clay usually formed by decomposition of feldspar minerals.

Karst: a distinctive type of landscape where solution in limestone layers has caused abundant caves, sink holes, and solution valleys, often with red soil residue.

Kerogen: solid oil-like substance in oil shales.

Laccolith: a body of intrusive rock that squeezed between rock layers, doming those above.

Lateral moraine: an elongate moraine along the side of a valley, deposited at the side of a valley glacier.

Latite: volcanic rock particularly high in feldspar minerals.

Leached: depleted of elements and minerals by slowly moving water.

Lode: a deposit of valuable minerals; a vein in solid rock, in contrast to placer deposits.

Magma: molten rock from which igneous rocks eventually solidify.

Magnetite: a magnetic ore of iron (Fe_3O_4).

Marble: a metamorphic rock created by baking and recrystallization of limestone.

Massif: a mountainous mass that has relatively uniform geologic characteristics.

Mesa: a tableland or flat-topped mountain or hill, usually capped by a resistant rock layer and edged with steep cliffs.

Metamorphic rock: rock formed from older rocks that have been subjected to great heat and pressure or to chemical changes.

Mica: a group of minerals characterized by their way of separating into thin, platy flakes, usually with shiny surfaces.

Migmatite: see *Injection gneiss.*

Monadnock: a hill or mountain remaining above a land surface that has been reduced by erosion to an almost level plain.

Monocline: a fold or flexure in stratified rock in which all the strata dip in the same direction.

Monzonite: an intrusive rock containing mostly feldspar minerals, with very little quartz and few dark minerals.

Moraine: an accumulation of gravel, boulders, and dirt, deposited by a glacier.

Muscovite: white or light brown mica.

Nunatak: an isolated hill or peak that sticks through a glacier.

Oil shale: shale containing kerogen, a waxy substance from which petroleum may be obtained by heating.

Ore: a rock high enough in valuable minerals to be economically mineable.

Orogeny: mountain-building.

Outcrop: bedrock exposed at the surface.

Outwash: gravel and sand carried from glaciers by meltwater and deposited below the actual glaciated area.

Oxbow lake: a crescent-shaped lake formed in an abandoned river bend.

Paleontology: a branch of geology that deals with plant and animal remains and the life of the past.

Paternoster lakes: a series of small lakes in a glacially eroded valley.

Patterned ground: rock circles, polygons, stripes, and other patterns due to frost action.

Pediment: a gently inclined erosion surface carved in bedrock at the base of a mountain or mountain range, most commonly in desert or near-desert environments.

Pegmatite: very coarse-grained igneous rock, usually occurring in veins, like granite in mineral composition.

Peneplain: land surface worn down by erosion to a nearly flat plain.

Periglacial: around or peripheral to a glacier.

Phenocrysts: large conspicuous crystals in finer igneous rock.

Placer: a gravel or sand deposit containing particles of gold or other valuable minerals.

Plateau: a large relatively high area limited on at least one side by cliffs or steep slopes.

Pyrite: a metallic, brass-colored iron ore mineral (FeS_2) often called "fool's gold," used as a source of sulfur for sulfuric acid.

Pyroxene: a group of dark minerals common in igneous rock.

Quartz: a hard, glassy mineral, silicon dioxide (SiO_2) that is one of the commonest rock-forming minerals.

Quartzite: a metamorphic rock formed from sandstone cemented by silica.

Recessional moraine: a glacial moraine formed during a temporary pause in the retreat of a glacier.

Redbeds: red, pink, and purple sedimentary rocks, usually sandstone and shale.

Rhyolite: light-colored volcanic rock containing invisibly small crystals of quartz, feldspar, and biotite.

Rock glacier: a glacier-like tongue of angular broken rock, usually lubricated by water and ice and moving slowly like a true glacier.

Rotten granite: granite weakened by decomposition of mica grains.

Sandstone: sedimentary rock composed of sand grains.

Scarp: cliff or steep slope.

Schist: metamorphic rock whose parallel orientation of abundant mica flakes causes it to break easily along parallel planes.

Shale: platy sedimentary rock formed from mud or clay, breaking easily parallel to the bedding.

Shear fault: a break in rock along which horizontal movement has taken place (more properly a strike-slip fault).

Silica: silicon dioxide (SiO_2), occurring as quartz and as a major part of many other minerals.

Siliceous sinter: hotspring deposits composed largely of silica.

Silicified: impregnated with silica.

Sill: a tabular sheet of igneous rock formed by hardening of magma intruded between horizontal or nearly horizontal rock layers.

Sinkhole: a large depression caused by collapse of the ground into an underlying limestone cavern.

Slickenside: a scratched and polished surface resulting from movement of rock against rock along a fault.

Stalactite: a calcium carbonate "icicle" hanging from the roof of a limestone cave.

Stalagmite: a columnar or ridge-like deposit of calcium carbonate rising from the floor of a limestone cave.

Stock: a medium-sized mass of intrusive igneous rock, smaller than 40 square miles at the surface.

Strata: layers or beds of rock. Singular is stratum.

Stratified: formed in layers,. as sedimentary rock.

Syncline: a downward fold in layered rocks. When eroded a syncline has the youngest rocks in the center.

Tailings: waste debris from ore-processing mills.

Talus: fallen broken rock collected at the foot of a hill or cliff.

Telluride: a compound of the element tellurium, often an ore of gold.

Terminal moraine: bouldery glacial debris dumped at the lower end of a glacier at the time of its greatest extent.

Thrust fault: a fault in which one side is pushed up and over the other side.

Travertine: hotspring deposits composed largely of calcite.

Tuff: a rock formed of compacted volcanic ash and cinder.

Type locality: the place at which a formation is best or most typically displayed, and from which it is named.

Unconformity: a surface of erosion that separates younger strata from older rocks.

Vein: a rock containing ore minerals, usually a tabular mass.

Volcanic ash: fine material ejected into the air from a volcano.

Water table: the upper surface of groundwater, below which soil and rock are saturated.

Welded tuff: volcanic ash which has been hardened by the original heat of the particles and the enveloping hot gases.

Xenolith: a large piece of other rock included in igneous rock, such as schist in granite.

Index

Cimmaron Ridge, 227
cinder cone, 318
cirque, 112, 113, 318
Clear Creek, 88, 115, 153, 252
Clear Creek Canyon, 88
Clear Creek graben, 252
Cliffhouse Sandstone, 293, 294, 296, 297, 298
Cliff Palace, 298
Climax, 8, 144
clinker beds, 131
clinkstone, 101
closure, 41
coal, 41, 131, 166, 167, 239, 260, 263, 270, 272, 273, 275, 292, 300, 301, 302
Coal Creek, 241
Coaldale, 68, 163, 165, 166
Collegiate Peaks, 108, 138, 214
Colorado City, 32, 34
Colorado Fuel and Iron Works, 32, 34, 171
Colorado (lowest point) 61
Colorado, map of, xiv
Colorado Mineral Belt, 82, 85, 87, 111, 115, 171-173
Colorado Mountain College, 204
Colorado National Monument, 259, 264-266, 268, 269, 285, 287
Colorado Orogeny, xiii, 105
Colorado Piedmont, 6, 20, 21, 23, 24, 33, 46, 47, 120
Colorado Plateau, 254, 255, 256, 265, 284
Colorado Ranges, 108
Colorado River, 6, 122, 123, 125, 126, 127, 130, 152, 154, 156, 157, 202, 203, 213, 255, 258, 261, 262-265, 269, 284-288, 303, 307
Colorado Rockies, 108
Colorado School of Mines, 71, 82
Colorado School of Mines Museum, 8, 71, 82, 116
Colorado Springs, 7, 28-34, 57, 58, 60, 70, 88, 89, 99, 110, 132-134, 139, 215, 216
columnar joints, 88, 318
Comanche Creek, 47
concretions, 273, 318
conglomerate, 13, 318
Conifer, 182-185
conodonts, 94
Continental Divide, 6
Conundrum Creek, 209
copper, 9
Copper Mountain, 121, 122
corn, 61, 65
Cortez, 221, 243, 244, 246, 248, 293-295, 298
Cotopaxi, 163, 165
Craig, 270-276, 299, 300
Craig Dome, 273
crater, 96
Crater Lake, 202, 209
craton, 3
Creede, 8, 248-252
Creede Volcano, 249-251

Cresson Mine, 100
Cretaceous Period, xii
Cripple Creek, 7, 71, 89, 99-101
Crook, 50, 51
Cross Mountain, 274, 276
crossbedding, 14, 75, 83, 92, 290, 308, 318
Crystal River, 204
Crystal Peak, 135
Cucharas Pass, 40
Cucharas River, 36
cuesta, 35, 255, 318
Culebra Creek, 195
Culebra Mountains, 108
Cumbres Pass, 198
Cutler Formation, 291
cyclic bedding (in coal), 166

dacite, 220, 318
Dakota Sandstone, 32-35, 39, 40, 61-63, 66, 68, 69, 77, 83, 116, 117, 119, 122, 124, 128, 130, 131, 145, 146, 154, 156, 160-162, 165, 170, 173, 175, 177, 183, 186, 200, 215-217, 224, 227-229, 231-234, 241-243, 246, 248, 256, 260, 264-266, 268, 270-272, 274, 279, 280, 285-291, 293-296, 300, 303, 304, 308, 309, 312, 313, 315
Dakota Hogback, 22, 35, 38, 39, 64, 74, 76-78, 81, 83, 115, 116, 120, 131, 135, 145, 156, 159, 161, 162, 182-186, 200, 215, 274, 277, 278, 280, 292, 302
Darton, N.H., 65
Datil volcanics, 107
daughter products, 11
Dawson Butte, 72
Dawson Formation, 18, 25, 27, 28, 30, 47, 48, 57, 58, 60
Deadwood Mine, 100
Debeque, 259, 262, 263
Debeque Canyon, 259, 263
decomposition, 112
Deep Creek, 122, 125
Deer Trail, 47, 48, 49
Del Norte, 228-230
Del Norte Peak, 229
Delta, 285, 286
dendritic drainage, 297, 298, 318
Denver, 7, 22, 24, 28, 30, 47, 48, 54-56, 70-73, 75, 81, 82, 110, 115, 120, 182, 184, 199, 201
Denver Basin, 20, 24, 25, 27, 34, 35, 47, 49, 55, 56, 63, 65, 68, 72, 81, 85
Denver Formation, 18, 24, 25, 27, 28, 30, 47, 48, 49, 54, 56, 58, 72, 82, 83, 182
Denver Museum of Natural History, 8, 26, 71
desert pavement, 275, 295, 318
desert varnish, 275, 295, 318
Devil's Backbone, 145, 146
Devil's Slide, 102
Devonian Period, xii
differential weathering, 76
dike, 36, 42, 64, 79, 314, 318

Holly, 62
honeycomb weathering, 263, 272, 301, 319
hoodoo, 59
Hoosier Pass, 100
hornblende, 15
hornblende gneiss, 116, 319
horse (early), 23, 174
Horseshoe Mountain cirque, 187, 188, 190
Horseshoe Park, 147, 149, 150
Horsetooth Reservoir, 200
horst, 177
Hot Sulphur Springs, 154-156
Howard, 165
Huerfano, 32, 37
Huerfano Butte, 32
Hugo, 44
Huron Peak, 212, 213
Hutton, James, 76
hydrostatic, 319
hydrothermal, 319
hydrothermal alteration, 193

Ice Age, xii, 57
ice cap, 113, 319
Iceberg Lake, 147
icthyosaurs, 46
Idaho Springs, 7, 78, 86-88, 115-118
Idaho Springs Schist, 86, 118, 162
Idarado Mine, 243, 316
igneous rocks, 12, 13, 319
Iliff, 50
incompetent, 319
Independence (town of), 99, 100
Independence Pass, 212-214
Indian Peaks, 25, 55, 108
Indian Paint Mines, 58, 59
injection gneiss, 165, 319
Inoceramus, 35, 46, 65, 158
insect fossils, 136, 138
interfinger, 39
intermittent stream, 61
I-70 road cut, 76, 116, 117
intertongue, 39
intrusive igneous rocks, 12, 13, 319
iron production, 34, 35
Iron Springs Bench, 280, 281
Ironclad Hill Mine, 100
Ironton, 236

Jackson, W.H. 250
Jackson Mountain, 289
James Peak, 98, 153
Jefferson, 182
Jefferson, President, 98
Jemez volcanics, 107
John Martin Reservoir, 62, 63
joint, 11, 78, 79, 319
Julesburg, 50
Juniper Mountain, 274, 276
Jurassic Period, xii

kaolin, 15, 192, 193, 319
karst, 129, 130, 239, 240, 319

Kawuneeche Valley, 152
Kayenta Formation, 260, 265-267, 308, 309
Keenesburg, 54, 66
Kenosha batholith, 184, 185
Kenosha Mountains, 108
Kenosha Pass, 80, 182, 184, 185
kerogen, 9, 319
kerosene, 9
Kiowa Creek, 47
Kiowa Peak, 83
Kirtland Formation, 291
Koshare Indian Museum, 66
Kremmling, 154, 155, 157-159

laccolith, 41, 42, 171, 294, 295, 319
LaJunta, 61, 62, 65, 66, 68
Lake City, 8, 252, 253
Lake City caldera, 249
Lake Creek, 140, 213, 214
Lake Estes, 148, 152
Lake Fork (Gunnison River), 249, 253
Lake George, 133, 135
Lake Granby, 152
Lake San Cristobal, 252
Lake Uinta, 257, 276
Lamar, 62
landslides, 315
LaPlata Mountains, 238, 244, 292, 293, 299
LaPlata Peak, 212-214
Laramide, 140, 141
Laramide Orogeny, xiii, 12, 78, 79, 81, 87,
    90, 96, 105, 109, 111, 141, 177, 196, 199,
    207, 214, 315
Laramide Range, 106, 200
Larkspur, 72
Las Animas, 62-64
Last Dollar Mine, 100
lateral moraine, 320
latite, 13, 220, 221, 320
LaVeta, 175
LaVeta Pass, 177
lava flows, 286
Law of Superposition, 11
leached, 320
lead, 8
Leadville, 8, 139-142, 144
Leadville Limestone, 128-130, 132, 143,
    167-169, 191, 205, 207, 212, 213, 235,
    239, 240, 276, 300, 303
Leavick, 187
Lee Mine, 100
lettuce, 65
Lewis Shale, 290, 291
Lime Creek, 234
Limestone, 13
Limon, 44, 45, 47, 48, 57, 58, 60
limonite, 15
Lincoln Gulch, 211, 213
Lincoln Memorial, 202, 205
Lindenmeir Site, 53
Lipalian Interval, xii, 85, 94, 95, 130, 169,
    201
Little Black Mountain, 188, 190

Ute Mountains (see Sleeping Ute
    Mountain)
Ute Pass, 88
Ute Pass Fault, 59, 89-91, 132, 134

Vail, 122, 123
Vail Pass, 122, 123
Valle Salada, 135
valley glacier, 113
vanadium, 9, 308, 310, 314
Vancorum, 310
varves, 257
veins, 79, 322
vesicles, 286
Victor, 89, 100, 101
Villa Grove, 194
Virginia Dale, 22, 200
volcanic ash, 137, 322
volcanic breccia, 219

Wagon Wheel Gap, 248-250, 252
Walcott, Charles, 162
Walsenburg, 32-34, 36, 38, 174-176
Ward, 82, 83, 85, 87
Wasatch Formation, 126, 128, 132,
    257-261, 264, 270, 272-274, 276, 277,
    284-287, 300, 302, 303
Wasatch Range, 55
water resource, 114
water rights, 49
water table, 322
weathering, 79
weathering, chemical, 112
weathering, of granite, 96, 101, 102, 112
weathering, of sandstone, 301
Weber Canyon, 298
Weber Formation, 274, 277, 278-281
welded tuff, 225, 226, 322
Wellington, 22, 24
Wellsville, 163, 165, 167
West Aspen Mountain, 207
West Creek, 287, 305
West Elk Breccia, 170, 171, 174, 179,
    223-227
West Elk Mountains, 108, 174, 219, 220,
    223-225, 253, 255, 275, 281, 282
West Maroon Creek, 209
West Monument Creek, 59

West Needle Mountains, 238, 241
West Peak, 39
West Plum Creek, 73
Western State College, 170
Westwater Canyon, 264
Wet Mountains, 32, 35, 36, 66, 68, 69, 107,
    108, 174, 175, 176, 217
Wetterhorn, 232, 233, 252
Wheeler Geologic Area, 248, 249
Whitehouse Mountain, 205
White River, 277, 302
White River Badlands (So. Dakota), 23
White River Formation, 22
White River Plateau, 108, 122, 125, 219,
    255, 273, 299, 301, 302, 303
White River Uplift, 2
Whitely Peak, 158, 159
Writewater, 285, 287, 301, 304, 305
Wiggins, 54, 56
Wilkerson Pass, 133, 135
Williams Canyon, 71, 93, 94, 134
Williams Fork Mountains, 270, 273
Williams Range, 120, 154
Williams Range Fault, 154
Wilson Peak, 245, 312, 313
wind blast pitting, 79
Wingate Sandstone, 242, 260, 264-269,
    287, 304, 308-311
Winter Park, 154, 155
Wisconsin Glaciation, 113
Wolcott, 122, 123, 125
Wolcott, Charles, 162
Wolf Creek Pass, 228-231
Wolf Mountain, 270
Wolford Mountain, 158, 159
Woodland Park, 89, 132-134
Woods Mountain, 153

xenolith, 172, 322

Yampa Plateau, 274, 276, 277, 279, 281
Yampa River, 6, 159, 161, 270, 274-281,
    300
Yellow Mountain, 245
Yule Marble Quarry, 202, 205

Zeolite, 286
zinc, 8, 142, 144
Zion National Park, 311

Check for our books at your local bookstore. Most stores will be happy to order any which they do not stock. We encourage you to patronize your local bookstore. Or order directly from us, either by mail, using the enclosed order form or our toll-free number, 1-800-234-5308, and putting your order on your Mastercard or Visa charge card. We will gladly send you a complete catalog upon request.

Questions? Call us toll-free, 1-800-234-5308.

Some other geology titles of interest:

____ROADSIDE GEOLOGY OF ALASKA	12.95
____ROADSIDE GEOLOGY OF ARIZONA	12.95
____ROADSIDE GEOLOGY OF COLORADO	11.95
____ROADSIDE GEOLOGY OF IDAHO	14.95
____ROADSIDE GEOLOGY OF MONTANA	12.95
____ROADSIDE GEOLOGY OF NEW MEXICO	11.95
____ROADSIDE GEOLOGY OF NEW YORK	12.95
____ROADSIDE GEOLOGY OF NORTHERN CALIFORNIA	11.95
____ROADSIDE GEOLOGY OF OREGON	11.95
____ROADSIDE GEOLOGY OF PENNSYLVANIA	12.95
____ROADSIDE GEOLOGY OF TEXAS	15.95
____ROADSIDE GEOLOGY OF UTAH	12.95
____ROADSIDE GEOLOGY OF VERMONT & NEW HAMPSHIRE	9.95
____ROADSIDE GEOLOGY OF VIRGINIA	9.95
____ROADSIDE GEOLOGY OF WASHINGTON	12.95
____ROADSIDE GEOLOGY OF WYOMING	11.95
____ROADSIDE GEOLOGY OF THE YELLOWSTONE COUNTRY	9.95
____AGENTS OF CHAOS	12.95
____FIRE MOUNTAINS OF THE WEST	15.95
____IMPRINTS OF TIME: THE ART OF GEOLOGY	19.95

Please include $2.00 per order to cover postage and handling.

Please send the books marked above. I enclosed $_____

Name————————————————————————————

Address ——————————————————————————

City ————————————————State ————— Zip ————

☐ Payment Enclosed (check or money order in U.S. funds)

Bill my: ☐ VISA ☐ MasterCard Expiration Date: ——————

Card No. ——————————————————————————

Signature ——————————————————————————

MOUNTAIN PRESS PUBLISHING COMPANY
P.O. Box 2399 • Missoula, MT 59806
☎ **Order Toll-Free 1-800-234-5308** ☎
Have your MasterCard or Visa ready.